Revealing

www.penguin.co.uk

Revealing

The Surprising Power
of Oversharing

Leslie John

torva

TRANSWORLD PUBLISHERS

UK | USA | Canada | Ireland | Australia
India | New Zealand | South Africa

Transworld is part of the Penguin Random House group of companies
whose addresses can be found at global.penguinrandomhouse.com.

Penguin Random House UK, One Embassy Gardens,
8 Viaduct Gardens, London SW11 7BW

penguin.co.uk

Penguin
Random House
UK

First published in Great Britain in 2026 by Torva
an imprint of Transworld Publishers

001

Copyright © Leslie K. John 2026

The moral right of the author has been asserted.

This edition published by arrangement with Avery, an imprint of Penguin
Publishing Group, division of Penguin Random House LLC

Image on page 28 based on image from Leslie K. John, Kate Barasz, and Michael I. Norton, "Hiding
Personal Information Reveals the Worst," *Proceedings of the National Academy of Sciences* 113, no. 4
(January 11, 2016): 954–959; chart on page 70 adapted from: David P. Schmitt, Jüri Allik, Robert
R. McCrae, and Verónica Benet-Martínez, "The Geographic Distribution of Big Five Personality
Traits: Patterns and Profiles of Human Self-Description Across 56 Nations," *Journal of Cross-Cultural
Psychology* 38, no. 2 (2007): 173-A, used with permission of David Schmitt; graphics on page 119
adapted from Leslie K. John, Alessandro Acquisti, and George Loewenstein, "Strangers on a Plane:
Context-Dependent Willingness to Divulge Sensitive Information," *Journal of Consumer Research* 37,
no. 5 (February 1, 2011): 858–873, used with permission of Oxford University Press.

Printed and bound in Great Britain by Clays Ltd, Elcograf S.p.A.

The authorized representative in the EEA is Penguin Random House Ireland,
Morrison Chambers, 32 Nassau Street, Dublin D02 YH68.

A CIP catalogue record for this book is available from the British Library.

ISBN:
9781911709299

For Dad

CONTENTS

Revealing

The Surprising Power of Opening Up

D o you overshare? Do you sometimes—possibly often— reveal more than you intended, unspooling your life story on a first date, before the breadsticks have even arrived? Or maybe you've shared your irritation about a boss with a colleague you have no reason to trust, or laughed loudly at a comment by a friend that you quickly discovered they meant quite seriously. If you're anything like me, the answer is a resounding, face-palmed "Yes."

If you're not an oversharer, you probably know one. Perhaps it's a friend who shared way too much about how good, or bad, the sex they were having was. (Really didn't need that image of them in your head.) Maybe it's a colleague who constantly updates you on the drama of his digestive tract. Or perhaps you've got a family member who regales one and all with his opinion on any manner of sensitive topics at get-togethers.

Oversharing takes many forms. Like getting personal too fast— think of the colleague who opens up about their personal woes during a meeting. Or talking so much about oneself that the person being unloaded to gets no chance to share as well—the "enough about me, let's talk about me" syndrome. Oversharing can also take the form of revealing to the wrong audience, at the wrong time, in the wrong way. Like a manager who publicly laments a tough decision that affected others far more than it affected them.

If you, like me, know all too well the sting of a disclosure hangover—that gut-wrenching sense of dread that you've said far too much about something utterly inappropriate—take heart. You're in good company. And not just because you share the experience with thousands of people who have participated in my studies, not to mention the vast majority of sentient human beings. I'm here to tell you that there can be a tremendous upside to revealing more than you think you should.

In fact, after almost two decades of studying when and why people disclose information about themselves, I've come to the conclusion that "oversharing" is underrated.

Often, when we are cringing with shame after an epic overshare, the person we've shared with is actually happy to have been confided in. My research, and that of others I'll introduce, has shown that undersharing deprives us of marvelous opportunities to build friendships, to gain the trust and support of colleagues and employees, and maybe even to find a life partner. Revealing more about ourselves than might feel comfortable is central to many of life's richest rewards, from the healthy rush of "happiness chemicals" released in our brains when we open up with people, to creating strong bonds of intimacy and feeling that we are truly understood. Yet so often we suppress ourselves.

Take John and Grace, who fell madly in love after a whirlwind romance. The only problem was that John had been visiting Grace in Washington, DC, on a student exchange, and so had to return to his native Australia. A few months later, Grace cashed in all her savings for a plane ticket to visit him. At the last minute, she had second thoughts. Had they grown apart? Would it be the same? Was she about to get her heart broken? Desperately seeking affirmation, Grace called John right before boarding. "Should I come?" she asked.

The best he could muster was a painfully noncommittal "If you want to." Grace didn't get on the plane, and they broke up.

In truth, both were overcome by fear and uncertainty. When Grace heard John's reply—flat, detached, and noncommittal—she panicked. What she had meant to ask was far more vulnerable: *Do you still love me? Are you still sure? Will I be safe if I come?* But she was terrified that asking directly would force an answer she wasn't ready to hear—that he had moved on, that she was a fool for chasing a dream across the globe. So instead, she hedged. She asked a question she thought was safer, and got an answer that only deepened her doubts.

John, for his part, genuinely wanted her to come. But he was also overwhelmed—young, emotionally inarticulate, and anxious about the responsibility her decision seemed to place on him. Rather than express what he felt, he fell back on something that felt neutral. He didn't want to pressure her. He wanted her to want it, too. But his attempt to sound casual came off as indifferent, and in trying to avoid saying the wrong thing, he failed to say anything meaningful at all.

They moved on with their lives, married others, and raised families, but both wondered what might have been. The years passed. But then John got divorced. And so, twenty years after he had last talked to Grace, he decided to track down her phone number. What was there to lose at this point, anyway? So he gave her a ring. On that short call, Grace mentioned that she had recently separated from her husband. And it was clear that she still had feelings for John. Soon thereafter, John and Grace began having video calls and were chatting for hours.

Before long, John visited Grace. Together they read the letters he had sent her decades before, which Grace hadn't been able to throw away. "It literally made us cry," she recounted, "to see the depth of emotion then, and that we could have it slip away."

This time, they didn't let it slip. What began as a tentative reconnection turned quickly into something undeniable. They spent days talking, laughing, crying—finally saying the things they hadn't been able to say in their twenties. Over the next year, they crossed oceans to see each other, video-called constantly, and slowly wove their lives back together. John joined Grace in the United States, and they've been inseparable ever since. They made a pact: Never again would they leave their feelings unspoken. They now realized that was what had torn them apart the first time.

If their feelings had been so strong, why did they both hold back when Grace made that fateful pre-boarding call? Their youth and physical distance from each other certainly made things harder. But they were also caught in one of life's most daunting disclosure dilemmas: whether to reveal deep feelings when you're unsure if they'll be reciprocated. So many of us get stuck in that guessing game. We monitor for signals, overanalyze tone, and make hesitant calculations about how much to reveal. We try to protect ourselves from rejection, often undermining the very intimacy we crave.

In this book, we'll explore why this happens—how universal cognitive traps like the *impact bias* (we overestimate how bad we will feel if things go wrong) and the *omission bias* (harmful actions seem worse than harmful inactions) push us toward silence. We'll also explore how personality plays a role—why some people, like Grace, default to self-censorship out of fear, while others, like John, freeze up under the weight of their own emotions.

Most important, we'll learn how to navigate these moments better. Because when we do speak—when we open up in the right ways, at the right times—what's possible can be extraordinary. Grace would later describe what made her love for John so enduring. "He loved me for me . . . all the flaws, without me having to do or be or achieve anything—just for me. And I think it's very rare to find that person that just knows you, that can truly see you and love what they see."

It is rare—but not unreachable. The rest of this book is about how to cultivate that kind of openness: not reckless oversharing, but thoughtful revealing. The kind that builds trust, deepens connection, and lets us be seen for who we really are.

But John and Grace's story isn't just about love—it's about what can happen when we don't say the thing. Every day, in all corners of life, we face decisions about whether to speak up or stay quiet—about how much of ourselves to reveal. Do you tell a friend that you're hurt they didn't include you? Do you tell your spouse that you're feeling distant, or your parent that you're worried they're no longer safe behind the wheel? Do you tell your boss you're struggling? Or that you think you deserve a raise? So often, we suppress what we're thinking and feeling, not because we have nothing to say but because we're afraid of saying the wrong thing, or of what might happen next.

The result is a persistent, under-recognized problem. It's not too much information, but too little. TLI. And so we all must face the fundamental question addressed by this book: how to navigate that gossamer-thin line between TMI and TLI. It's a challenge that even beauty queens, standing in sequins under hot lights, know all too well.

The Goldilocks Principle

The 1997 Miss Universe competition culminated, as it had for years and years, with the dreaded final question—well, dreaded from the contestants' standpoint; more like hotly anticipated for me. There's something about beauty pageants that fascinates me. I love the spectacle of it all.

For the final round of the competition, the three finalists each have to give a twenty-second answer to a hot-button question. There are no right answers, but plenty of wrong ones. Sometimes the questions are softballs, like "What is your definition of success?" But sometimes they're trickier, like when the finalists were asked what made them blush (no one did very well with that question) or when they were challenged to explain why the pageant isn't disrespectful of women—to which Miss India made such a powerful statement about the platform serving as a springboard to successful careers in politics, entrepreneurship, and the military that she was crowned Miss Universe 2000.

In the 1997 competition, the final three contestants are Miss Venezuela, Miss USA, and Miss Trinidad and Tobago. Marla Maples, Donald Trump's then-wife, who (fun fact) is cohosting the show this first year her husband is producing it, tees it up: "Now the tension really goes up a notch. Those three have one last round of competition—the final question." The other host, George Hamilton, his face practically orange from all the tanning and makeup, announces: "Ladies, I will ask each of you the same question. You'll have twenty seconds to respond."

Miss Venezuela, towering over the other two, is first up. With a gorgeous off-the-shoulder white gown, the odds are in her favor because (another fun fact) white is by far the most likely gown color to win. Two tux-clad men escort Miss USA and Miss Trinidad and Tobago into a transparent soundproof booth to keep them from hearing the question.

Hamilton reads from a card: "If there were no rules for one day, and you could be outrageous, what would you do?" Miss Venezuela promptly responds that she'd take the day to travel, magically fly-ing from place to place. Her turn complete, she's then whisked off to the side.

Next, the soundproof doors open and out slips Miss USA, Brook Mahealani Lee, in her spunky, slightly risqué, royal blue gown in all its faux midriff-baring glory. As Hamilton asks the question, her face lights up with excitement and joy. She immediately responds: "I would eat everything in the world!" This sends the audience, Ham-ilton included, into stitches. She continues, "You do not understand. I would eat everything twice." Uproarious laughter.

Finally the crowd quiets down, and Miss Trinidad and Tobago emerges from the booth in a glittering strapless gold gown, com-plete with a taffeta shawl. She looks nothing short of regal. Like the other contestants, she also answers without hesitation, but less jo-vially than Miss USA. "I would not wear clothes," she says in a strangely earnest tone. This did not land well. She tries valiantly to save her response by qualifying it: "Clothes . . . I mean, it's neces-sary, because we know we have to be private and all that." Digging her heels in, she continues, "But if we had no rules, I would want to be free, and I'm sure everybody else would." Nope. She ended up second runner-up (that's third place in beauty language parlance). And the winner? Miss USA, of course.

By and large, we like people who reveal to us. And by "reveal," I mean disclosing personal information that is somewhat sensitive or unexpected. Miss Venezuela's response was pretty superficial. Say-ing you want to travel is hardly something that makes us feel like we know you—TLI. Miss Trinidad and Tobago, on the other hand, went too far. Her response was a little provocative for this crowd—TMI.

By contrast, Goldilocks would have judged Miss USA "just right." Having watched my share of beauty pageants, I can tell you that the answers usually sound extremely scripted. But hers felt like a spontaneous, unfiltered blurt. It was refreshing, and funny, too. The admission that even beauty queens love to eat made her relatable and therefore likable. It also struck me that she was getting real about what it takes to prepare for a beauty contest and the self-discipline (read: deprivation) it entails.

Because I'm a nerd, I decided to collect a little bit of data on this. So I showed this clip of the final question to a couple hundred people as part of an experiment and asked them how authentic they perceived each finalist to be—as in how "real" they were, and also who they liked best. The findings of my oh-so-serious study were very clear: By far, people liked Miss USA the best, and this was largely because she came across as authentic (we will look more at self-disclosure and authenticity a little later in this book). Overall, reveals that hit the sweet spot between TMI and TLI breed liking, which is a precursor to so many of the best things in life, like friendship, love, and status.

But, as this story shows, nailing that sweet spot can be challenging (and being liked is not always the end goal). Context matters so much. Miss Trinidad and Tobago's reveal was TMI for a conservative beauty-pageant crowd, but it might have come across as excitingly candid in a looser environment. The same could be said of Miss USA. Though her reveal was just right for that moment, you can envision it backfiring in other contexts. A very thin person saying they'd eat everything might sound insensitive in a group where people are struggling with food or body image. There's an art to hitting the sweet spot of sharing, and as you'll soon see, there's also a whole lot of science.

I have largely spent my career asking people about the most taboo, off-putting things, and there are certain topics that are consistently sensitive across most—or I daresay even all—contexts. The first time I devised a survey about sensitive behaviors, I was in graduate school at Carnegie Mellon. Coming up with sensitive topics with my PhD adviser, the person I most wanted to impress at that point in time, was about as enjoyable as walking barefoot across a sea of small LEGO bricks. It didn't help that I'm a blusher.

"What about bestiality?" he began.

I froze. That was . . . a bold opener.

From there on out, I worried that everything I suggested for the list of sensitive topics would imply something about myself—that I'd come across as either prudish ("Have you ever masturbated?") or horribly irresponsible ("Have you ever neglected to tell a partner about an STD from which you were currently suffering?").

Nonetheless, I got through the brainstorming sesh, we collated our list of behaviors, and we asked people to rate how invasive it would be to be asked about each one. The results were extremely consistent: People generally deemed questions about sexual preferences and behaviors, personal finances, and immoral, unethical, or taboo behavior to be extremely sensitive and invasive. (The sensitivity of these items also happened to correlate with the redness of my face during the brainstorming session.) I've repeated this poll periodically, and the results have stayed consistent. They're also quite consistent across different cultures.

But making good disclosure decisions doesn't mean simply avoiding taboo topics. On the contrary, talking about sensitive things is where the huge opportunity—as well as the risk—lies. Hence how utterly conflictual and nerve-racking disclosure decision-making can be. The challenge is calibrating—not too much, not too little, but just right—the same lesson that a certain blond girl in a bear-owned house had to learn the hard way. As we'll see, making better decisions in this space requires upping both our self-awareness and our situational awareness.

Like the classic fairy tale, we'll be testing, adjusting, and refining. Just as Goldilocks had to go through multiple porridge bowls, chairs, and beds before finding the perfect fit, making good disclosure decisions takes practice. Getting to "just right" in disclosure isn't about rigid rules; it's about trial and error, context, and calibration. And about making our unconscious processes more conscious, so we know when and why we share or don't.

Why I Wrote This Book (and What You'll Get from It)

This book is the culmination of my work studying disclosure decisions for many years. It's about the choices we make every day—at work and at home, with our spouses, friends, colleagues, and even strangers—about what we reveal about ourselves and what we're thinking and feeling. The more I studied these decisions, the more intrigued I became. Because making them well can be so tricky, and the right choices are so often counterintuitive. But I also began to realize how central these decisions are to our well-being.

Sometimes we're confident about where to draw the line. We're talking to a close friend and have no doubt we can trust them with knowing that we're terrified about making a big presentation at work, or that we're going crazy because we haven't heard from someone we've been dating. Maybe we're sure that we can tell a close colleague that our boss has just told us we're on shaky ground. We know we'll get some good support and ideas about how we might turn the situation around. With a spouse or partner, we may feel secure telling them about some annoying habit they have. But often, deciding whether and how much to share is difficult, even excruciatingly so.

Many of our disclosure decisions are in fact disclosure *dilemmas*, anything but black-or-white. We are of two minds, simultaneously tempted to reveal and conceal. If you keep your mouth shut when people are trash-talking the boss, no one can accuse you of office gossip. But you also risk missing a chance to connect with your co-workers and gain their trust (not to mention process that weird meeting). Should you choose safety or solidarity? Should you divulge a medical diagnosis to your boss? Go to HR to report a colleague? Tell your partner about that old fling? Suggest to a friend that you are concerned that she's drinking too much?

We often fail to recognize that we should consider more carefully what we're revealing. We sometimes make unfortunate disclosures so rapidly that they're like physical reflexes on par with a hiccup or a sneeze. Consider how many times a day you get asked how you are. If you're having a bad day, you might shoot back, "I feel like shit" or even "What do you care?" You might come home after having a horrible meeting with a client, and when your partner asks how you are, immediately start unloading. While such rapid-fire sharing can be great for creating and strengthening bonds, it can sometimes also be a big turnoff. Even when we think we're sharing in order to provide support, we can misstep. Consider when someone confides in us and our well-intentioned impulse is to share our own struggles with the same issue, when instead our confidant just wants us to listen.

My research on disclosure decisions has illuminated just how difficult it can be to adjudicate between too much information and too little. One detail too many about your health problems, and your boss's concern may give way to questions about your competence. If you share that juicy bit of gossip about a coworker with another colleague, your confidant might be flattered and respond with a juicy morsel in turn—and, voilà, friends for life. Then again, the colleague you shared with might worry you'll gossip about them as well, leading them to keep their distance from you.

As my and others' research has revealed, all sorts of features of the situation you're in can influence your disclosure behavior, making disclosure decisions even more complicated. The gender and attractiveness of the person you're talking to, the power dynamics between you, whether you're communicating via cell phone or on a laptop, even the coziness of the room—all can affect self-disclosure.

I wrote this book because I've learned how important it is to get disclosure decisions right. How well we make them has a powerful effect on not only our daily happiness, the quality of our relationships, and our fortunes in our careers, but also our health. Sharing more freely can increase our immune functioning, reduce depression, and increase our overall sense of well-being. As we'll see, the physical and mental health benefits of revealing are not restricted to those already suffering from health problems. They are seen across broad swaths of people, young and old, educated and less educated, optimists and pessimists alike. Time and again, revealing has been found to improve our health and general life functioning.

Yet the difficulty of working through disclosure dilemmas, and our fear of the consequences if we get our decisions wrong, leads so many of us to err on the side of undersharing. Indeed, the notion that this is the better option has led to the grossly simplifying mantra, *When in doubt, say nothing.* So many of us have come to see what would be beneficial disclosure as TMI. I wanted to know how we can do a better job of disclosing. How can we learn to disclose more when doing so would be beneficial, while also being mindful about when it truly would be better to stay mum?

First, I realized, we need to become more aware of the lengths we often go to in order not to disclose, as though we're playing a concealing game, like high-stakes poker. We can take a very different approach, I'll show, which will allow us to safely reap the rewards of sharing. In what follows, we'll explore the dangers of concealing. I'll also help you get a sense for your personal set point: Are you someone who tends to be reserved and keep things close to the vest? Or do you tend to let it all out? Either way, this self-knowledge is key to adroitly navigating all the disclosure dilemmas life throws your way. In the second half of this book, we'll explore how nailing that sweet spot between TMI and TLI fosters many of life's richest rewards, like well-being, friendship, love, and influence. We'll learn how to capitalize on disclosure dilemmas in a well-calibrated fashion, keeping the very real risks of sharing *and* withholding top of mind.

My hope is that you'll leave this book equipped with the awareness to make disclosure decisions with aplomb and even a sense of adventure. To be clear, I'm not offering a laundry list of dos and don'ts. Disclosure decisions are simply far too complex for that. Instead, I'll offer you the latest scientific insights about how to hone your self-awareness and situational awareness so that in any given moment, you can make a good decision about just how much you want to share and why.

Along the way, I'll share stories of people navigating some of life's thorniest disclosure dilemmas (with a few names and details changed on occasion). We'll have fun exploring the answers to a host of intriguing questions I'm constantly asked: Do people with certain personality traits, like extraversion, share more than others? Is it true that women open up more than men? If so, why? Does talk therapy work? Is crying at the office career suicide? (Stay tuned for my own story on that one.)

There are a few questions in particular that I hope you'll ask yourself. Should I share more? With whom? When? And above all: What might I have been missing by holding back?

2

Why We Stay Silent

n his 1970s campus novel *Changing Places*, David Lodge introduced readers to a literary parlor game called Humiliation. The name of this game is also its object. Players—in this case, professors of literature—attempt to outdo one another by naming a classic work they've never read (but assume others have), scoring a point for every person who has read it. In doing so, Humiliation inverts the rules that govern these lit scholars' professional lives, where the goal is to have mastered the entire literary canon, or at least give that impression. Suddenly the most humiliating gap in their reading repertoire is the winning ticket. Stepping back, we can also see that Humiliation creates an ingenious double bind. Players win by losing and lose by winning. The urge to "win" collides with the fear of being thought uncultured, leaving the professors caught in a deliciously awkward double bind.

At first, Howard Ringbaum, Lodge's most obnoxiously pedantic and competitive character, is so focused on professional prizes like prestige and promotion that he is unable to enter the spirit of the game. Instead, he resorts to humblebragging, naming an obscure eighteenth-century book. His strategy is clear: Rather than risk real vulnerability, he picked a book so niche it was more likely to make his colleagues feel underread than to actually score him points.

Sulking after losing the first round, Ringbaum sat out the second round. During the third, his killer instinct shifted focus. He "slammed his fist on the table, jutted his jaw about six feet over the table and said: *'Hamlet!'*" With this one showstopping admission—*Hamlet*, possibly the most canonical text of Western literature next to the Bible—Ringbaum wins the game. And loses his job: The English Department "dared not give tenure to a man who publicly admitted to not having read *Hamlet*."

As comic and absurd as it is, this classic scene from *Changing Places* reminds us of the real risks of revealing—and why English professors would understandably be wise to keep notable gaps in their knowledge secret. (I sure kept my limited business experience under wraps on the first day I taught MBAs at Harvard Business School.) Letting others have privileged knowledge about us gives them power.

What Are Your Thirteen Secrets?

Michael Slepian, a professor of leadership at Columbia University, specializes in the study of secrecy. Early in his career, he set out to understand a very intriguing question—just how many secrets we typically keep, along with what kinds of things we choose to hide. He and his colleagues Jinseok Chun and Malia Mason developed the Common Secrets Questionnaire, one of the most comprehensive tools for studying secrecy. They built the questionnaire by first asking one thousand people to anonymously describe a secret they were holding. They distilled those responses into thirty-eight categories, including sexual behavior, infidelity, lying, self-harm, drug use, poor work performance, unpopular or taboo beliefs and preferences, and ambitions.

Since then, Slepian, Chun, and Mason have administered the full questionnaire to thousands of people from all walks of life. They have found that at any given moment people keep, on average, thirteen secrets. The most common secrets, perhaps unsurprisingly, are about our romantic lives. For instance, many of us keep secret what turns us on and who we are attracted to. We keep all sorts of other secrets, too. We don't tell our friends that we disapprove of their significant others. We're ashamed to tell them when we're struggling with our finances or fighting with our partners. Other secrets are more about personal preferences and social pressure, such as liking something considered "lowbrow" (such as reality TV or gossip magazines, both of which I enjoy) or disliking something we're "supposed" to enjoy. We keep loads of secrets in the workplace, too, like our political opinions and mental health struggles. Secret keeping is such a natural impulse that we often turn to it largely by default, with little deliberation.

A few years into my academic career, some faculty at UC Berkeley invited me out to give a talk, which was a thinly veiled job interview. Needless to say, I was on my best behavior. But a challenge emerged. Two of the most respected professors—an academic power couple—invited me to their home for dinner. I was flattered and excited. When we sat down at the table, my heart sank. The first dish they served was steak tartare. Raw beef, no matter how well seasoned, no matter how swanky the cut, is not my jam; nor is any raw animal matter, for that matter. And in the home of these illustrious professors, I was ashamed about that. I feared they would consider me unsophisticated for not liking the dish.

It's at these tough junctures in life that I turn to Mr. Bean for inspiration. In a flash I recalled one of my favorite sketches, in which he unwittingly orders steak tartare at a fancy restaurant. After the waiter lifts a silver cloche from the plate laid before him with a flourish, the aghast Mr. Bean frantically finds places to hide the offending beef away. He hollows out a dinner roll and crams some in; he drops a big dollop onto the table and plops the side plate onto it; he squishes the remainder into the flower vase just before the waiter returns to ask if everything is all right.

I couldn't, of course, do the same. But, like Mr. Bean, I concealed my revulsion. Pretending to be very excited at the special dish, I took as many bites as I could, emitting the requisite coos, while trying desperately to suppress my gag reflex.

Why didn't I just apologize and tell my hosts I don't like steak tartare? Surely they would have understood. They probably would have even apologized themselves for serving something I didn't like, and maybe offered me something else instead. But in that moment, in that situation, the stakes seemed incredibly high. I didn't want to come across as unrefined. I was trying my darnedest to fit the mold of "esteemed academic," especially because, being young and female, the deck was stacked against me. I thought that required acting worldly and sophisticated by liking high-class fare.

We often act impulsively when keeping secrets or turning what might be understandable disclosures into secrets. I did this myself when I put on my raw carnivore act. But our decisions to conceal aren't purely gut reactions. They're the product of powerful cultural conditioning. From a very early age, we are taught that we should carefully manage how we present ourselves to others. This ongoing "performance" was the focus of one of the most influential social scientists of the twentieth century. And it sheds light on why we withhold information and wrap ourselves in secrets.

All the World's a Stage

Meet Erving. He's a bit of an enigma. His friends called him "bright but strange." His full name is Erving Manuel Goffman, and he's best known for his book *The Presentation of Self in Everyday Life*. In it, he introduced the core idea that much of human behavior is performative, enacted on the "front stage," where we carefully control what we say and do to manipulate others' impressions of us. It's only when we're backstage—when no one else is watching—that we are truly ourselves, he conjectured.

Goffman honed these ideas as a doctoral student on the remote Shetland Islands of Scotland. Day after day, he watched how people seemed to withhold their true feelings. A mother sent her son off to sea "without kissing him and with very little show of emotion" though he'd be gone for years. A housewife burned her finger on a hot pot and "very little emotional expression [was] allowed to escape." Even in the small frustrations of daily life, concealment was the rule. And on the whole, Goffman approved.

That stance makes sense when you consider Goffman's background. Born in 1922 to a Jewish Ukrainian family in Canada, he avoided the Holocaust but not antisemitism. In 1939, Canada itself had turned away a ship of Jewish refugees, many of whom later perished in death camps. For someone raised in a lower-middle-class, stigmatized community, holding back seemed quite sensible.

It was from this vantage point that Goffman saw the world. By all accounts, he lived the restraint he studied. He banned recordings of his lectures, avoided photos, gave only two interviews, and even sealed his archives before his death. Stories abound of his stoicism. At his famously grueling PhD defense, legend has it a bead of sweat slid down his brow, and he didn't even flinch.

In short, Goffman saw careful self-preservation as armor: protection against embarrassment, rejection, and worse. He joined a long tradition—from ancient proverbs to modern psychology—arguing that sometimes, silence is a strength.

But self-protection isn't only about staying quiet. Sometimes we go on offense, bluffing to manage how we're seen. We say we couldn't care less that someone insulted or embarrassed us, when in fact we're steaming mad. We claim we didn't really want a job we weren't hired for, when in fact we're devastated. We also bluff sometimes because we think we'll make a better impression on others. If we're stepping into a new leadership role, we might put on a great show of being super confident, when in fact we're feeling anxiety about how well we'll do the job. On a first date, we might work fervently to come across as a lighthearted extravert when really we're a serious-minded, brooding introvert.

This bluffing of ours was fascinating to the brilliant mathematician and polymath John von Neumann, widely credited as the founder of the field of game theory. Whereas Goffman's interest in self-presentation was rooted in his childhood experiences, a large source of inspiration for von Neumann was his poker-playing habit (though rumor has it he was not a very good player). "Real life consists of bluffing, of little tactics of deception, of asking yourself what is the other man going to think I mean to do," wrote von Neumann. He was so taken with the notion that life is like a game of poker that he conducted in-depth analyses of the best poker strategies. He sought to work out a detailed game plan for when and how we should conceal and bluff in life.

Game theory helps us navigate strategic decisions—especially when the stakes are high, outcomes are uncertain, and success depends on predicting what others will do.

When it comes to our everyday interactions with others, we do sometimes approach them as though we're game theorists. We strategize about what to share and why, even if we don't always realize we're doing it. We're uncertain how others will react or whether they might use our disclosures against us. Will they think less of us? Will they be offended and shun us? Might they use the information to humiliate us? We're inclined to engage in this gamesmanship because we're quite good at bluffing and concealing.

And sometimes bluffing goes beyond saving face. We hide things that feel safer to keep inside, but that silence can put us at real risk.

Consider a shocking study that estimated about 80 percent of Americans have lied about health issues to their doctors—the very people who have taken an oath to care for them! This is often because we feel shame about our condition or behavior, such as drinking or smoking, and we shudder at the thought of being judged by someone we hold in high esteem. Those feelings are valid, but failing to overcome them can be life-threatening.

American Academy of Family Physicians president Dr. John Cullen recalled a time when he was literally about to cut open a patient to take out their appendix, but he had a hunch that something was not quite right. Apparently, symptoms of drug abuse (specifically, methamphetamines) can resemble those of appendicitis. Fortunately for the patient, Dr. Cullen gave it one final chance: "We're about to cut you open here. Are you sure you don't want to tell me anything else?" The patient fessed up, and the surgery was halted.

Another surgeon recounted a case of a woman who started to bleed uncontrollably midway through surgery. Prior to surgery, she had affirmed that she wasn't taking any medications. The nurse rushed to the waiting room to ask the woman's mother whether she was taking anything. Again, the answer was nothing. Finally, the surgeon asked, imploring the mother to be honest. The mom finally revealed that her daughter was taking weight loss supplements but had been too ashamed to say so. The surgeon ran back in, and a life was saved.

Not all concealment and life-or-death. Sometimes it's just about saving face—but even then, the costs can be real. Have you ever experienced a variation on the Humiliation game that I opened this chapter with? Situations in which revealing your greatest weakness could enable an immediate win—and also threaten a long-term loss? I have, more than once. I suspect most of us have. One memorable example for me took place when I was a doctoral student, at a conference party that went late into the night. I found myself sitting on the floor in a circle with a few hugely influential behavioral economists and several other senior academics in my field, none of whom had known of my existence beforehand. Somehow we got onto the topic of embarrassing stories. Everyone took their turn, à la Humiliation, mostly admitting to an array of not-so-embarrassing things while cleverly using the opportunity to work in some humblebraggy professional reveals—like finding a typo in the abstract of the article they had just published in a fancy journal (the horror!). I, like Howard Ringbaum, turned out to be the showstopper when I shared my most (*actually*) embarrassing story.

I was acting in a German play in college, *Der Besuch der alten Dame* (*The Visit of the Old Lady*) by playwright Friedrich Dürrenmatt. I played the strict schoolteacher who lets loose and gets very drunk in one scene. Well, I got *super* into it. The audience roared with laughter. I couldn't contain myself and also started laughing. Uncontrollably. So much so that I peed myself, onstage. In front of five hundred people. Worse, I was wearing a dress with thin pantyhose. Fearing the crowd could see the trickle of pee, which felt like a waterfall, I panicked and started to splash and dump the bottle of "vodka" (actually water) everywhere onstage. Until this moment, apart from that conference party, I've only told a couple of close friends about this. My family was in the audience, and we have never spoken of it. I don't know whether I successfully covered it up. (Recently, my mom gave me a photo from the play that she had found. I could be reading into things, but I swear I saw a wry smile cross her face. So maybe the jig is up? Well, it most certainly is now!)

As I wrapped up the story, I glanced around the circle. People were howling. But as the laughter ebbed, I felt a flicker of panic. Had I just gone full overshare in front of the very people whose respect I hoped to earn? Was this a career-torpedoing disclosure dressed up as comic relief?

The next morning, I had the biggest disclosure hangover of all time, the emotional equivalent of having drunk a bottle of cheap wine. Sure, I had been a hit—mine was definitely the funniest story of the night. But at what cost? How could I have been so stupid? It was like I had been on a stage and humiliated myself all over again.

Clearly, in the performance of life, I was one of the bad poker players.

Exposing Your Belly

After the first wave of game theory pioneered by John von Neumann, which focused on games of competition, economists were eager to extend this way of thinking into other realms—in particular, one of the great conundrums studied by economists, known as the Tragedy of the Commons. In a classic article that coined the problem, ecologist Garrett Hardin described common-pool resource problems, or situations where people must share the fruits of a scarce resource, such as a forest or a lake or the world as a whole. Each member accessing the "commons" has an incentive to take as much of the resource as they can to maximize their riches. But if they do so, they will keep the resource from replenishing itself and inevitably deplete it, leaving them all with nothing. Tragic, right?

As an example, take the case of fishing or, rather, *over*fishing. Individuals are incented, in the short term, to each take in as large a haul as possible. But if they all do so, they'll end up overfishing, and eventually there will be no fish for anyone.

Knowing about this tragic flaw in human nature, I felt like I was watching a real-life game theory experiment as I sat perched on a rock on the rugged coastline of Maine's Acadia National Park on a blustery fall day. I was hunkered down with my husband Colin and our baby boy Oliver for a nursing break during a hike. As I gazed out over the vast seascape, listening to the waves crash onto the shore and the wind howling, a modest boat appeared, operated by a single fisherman. He maneuvered over to a cluster of brightly colored buoys. This was lobster territory, and the buoys, each painted with distinctive colors—one bright blue, another red with a white stripe, and a third vibrant orange—were tethered to traps. The colors indicated that each trap was the property of a different owner.

The fisherman, who didn't see us, arrived at the blue buoy and pulled up the trap, revealing a bounteous haul. After emptying it, he lowered it back in and puttered away. Having been infected by Hardin's gloomy predictions in grad school, I watched in suspense, half expecting the fisherman to empty the other traps, too. It would have been so easy, and with no consequences. Sure, he would have had a secret he carried with him, but as we already know, we all carry secrets. The owners of the other traps would never know that lobsters had awaited them that particular day, or at least who had taken them.

That he did not raid the other traps was uplifting to me, and also sent me into a delightful research spiral to understand why. After all, he seemed to be a walking—er, floating—contradiction of the

Tragedy of the Commons. Plus, as dark a view as one might have of human nature, we all know that people often join forces in cooperative efforts, eschewing individual advantages for collective benefits. So what was going on?

Beginning in the 1960s, in fieldwork all around the globe, influential political economist Elinor Ostrom documented cases of people coming together to solve common-pool resource problems. She observed Maasai herders in Kenya working together to devise intricate systems that kept them prosperous but prevented overharvesting. Nepalese farmers figured out how to prevent deforestation while making a good living. And, yes, she also studied the lobstermen of Maine. In Ostrom's seminal book, *Governing the Commons*, she compared the outcomes of government-mandated cooperation—such as laws prohibiting farmers from cutting down trees to plant crops in certain areas of the rainforest—to the results of systems of cooperation devised granularly by stakeholders themselves. She found again and again that the local solutions worked best. Ostrom's discovery rocked the economics world and ultimately earned her a Nobel Prize.

In studying just how it was that people reached these self-organized solutions, Ostrom found that people most certainly did *not* get there by concealing. On the contrary, they got there by being open. By talking face-to-face and putting all their cards on the table. Being transparent about the problem at hand, about the temptation to be selfish, and about how the resources ought to be equitably divided. In other words, by being extremely reveal-y, as if collectively deciding to dispense with secrets in this one vital category of their lives. In turn, the communities were able to agree upon norms of use, allocation rules, and appropriate sanctions for misbehavior.

The Maine fisherman I had observed was clearly cooperating to assure both that he and his fellow lobster-hunters each got their fair share of lobsters and that lobsters would remain plentiful. They had agreed to each sink only one trap in a high-traffic zone. Classical economic theory long emphasized self-interest as the engine of human behavior. But in practice, we're often surprisingly cooperative—even willing to act for the good of others at personal cost. And these positive norms come to be through a shared process of disclosure.

Since Ostrom, many others have studied these emergent solutions to the Tragedy of the Commons. One especially hopeful finding comes from psychologists Rebecca Koomen and Esther Herrmann, who studied how six-year-old children handled a common-pool resource dilemma. They designed an elegant experiment to mimic the real-world temptation to overharvest: Pairs of children were given access to a limited but renewable resource—water—used to float plastic eggs that earned them candy rewards. If both children waited and coordinated how much water they used, the water would replenish, and they could maximize their haul. But if either took too much too quickly, the system collapsed—cutting off the water supply for both.

Some children developed cooperative strategies to prevent collapse: taking turns, proposing shared rules, and coordinating their actions. What set them apart was how openly they talked through the dilemma. Rather than staying silent or guessing what the other might do, they made their thinking visible, aligned their efforts, and solved the problem together.

Reading the accounts of discussions that led to solutions, I was struck by how they demonstrate something I've found in my own research—that revealing sensitive information about ourselves fosters trust and doesn't necessarily lead to people taking advantage of us. In Ostrom's research, as well as Koomen and Herrmann's, people had to talk about touchy subjects, such as how much more money some herders or fishermen were making than others. Yes, this is hard. But being willing to disclose such sensitive information about ourselves is an expression of trust in those we share with, and that in turn inspires them to trust us. My research shows that the reverse is also true. When we conceal, we *undermine* the trust people have in us and therefore short-circuit the potential for getting into a positive spiral of mutual sharing.

My collaborators, Kate Barasz and Mike Norton, and I discovered this in an experiment inspired by *The Dating Game*—that show where you choose a suitor based only on their answers to questions. In our version, participants saw how two prospective dates had (supposedly) answered questions on their dating profiles. Then we asked: Based on this alone, which suitor would you rather date?

But these weren't your typical profile questions—they were questions that make "What's your favorite color?" feel like a lost golden age of courtship. They were questions like "Have you ever made a false insurance claim?" We used such outrageous questions because we wanted to know whether concealing was so off-putting that people would actually prefer someone who openly admits to terrible things over someone who simply withholds their answer. So we varied how the two (fictional) suitors responded. As shown in the figure, which is akin to what participants saw, Eligible Dater #1 endorsed the most egregious response: Frequently. Eligible Dater #2 opted out of answering entirely. Neither candidate is exactly a fine specimen, but if push came to shove—if you had to choose one—whom would you pick?

Time and again—across different creepy questions, different response options, and different participant samples—we found that more than half chose Dater #1, the confessor. At first, I was floored by this result. After all, we don't know that Dater #2 did anything wrong; maybe they just found the question offensive.

Eligible Dater #1's answer:	Eligible Dater #2's answer:
Have you ever made a false insurance claim?	**Have you ever made a false insurance claim?**
○ Never or Once	○ Never or Once
○ Sometimes	○ Sometimes
● Frequently	○ Frequently
○ Choose not to answer	● Choose not to answer

We did more studies. We kept seeing the same pattern, sometimes even more strongly. Eighty-nine percent of participants preferred to hire someone who admitted to failing an exam over someone who withheld their grades. Eighty-five percent were more interested in a colleague who kept a painfully transparent calendar—including therapy and colonoscopy (!) appointments—than one who kept their entries private.

Why do people favor the confessor over the hider? What explains this seemingly irrational decision-making? Let's look to game theory and experimental economics, which rely on decision-making games to learn about people. These are stylized interactions where players face strategic decisions in pursuit of maximizing their payout. The Prisoner's Dilemma. Battle of the Sexes. Chicken. Stag Hunt. Cake Cutting. Princess and Monsters. The Muddy Children Puzzle. (I'm not making these up.)

For this puzzle, we turned to the classic Trust Game, where you decide how much money to hand over to a stranger, knowing you'll only profit if they prove trustworthy. Economists love it because there's real financial skin in the game. If you misjudge your partner, you lose money. (Economists don't ask whether you trust someone;

they watch what you do when cash is on the line. I still can't decide if it's ironic or fitting that a famously distrustful discipline invented a popular measure of trust.)

In our version, participants saw a profile supposedly filled out by their partner—either a revealer, who confessed to bad behavior, or a hider, who refused to answer. That was all the information they had. And over and over, participants entrusted more money to the revealers, even when the confessions were seriously damning.

As I considered why disclosing sensitive information about ourselves builds trust with others, I realized that vulnerability is key. When we make ourselves vulnerable by sharing information that could cause us harm, we're demonstrating trust in the people to whom we're disclosing. They, in turn, are inspired to trust us more. This isn't just a quirk of human psychology—it's a pattern seen throughout the animal kingdom. Examples abound, but let's take puppies. As dog people know, puppies are in the habit of exposing their necks or bellies to potential playmates. These are very vulnerable parts of their bodies. Otters, octopuses, and a host of other creatures, large and small, do versions of the same. Well, a human analogue of exposing your belly is being the first to reveal something potentially embarrassing or unflattering. Like how you once peed yourself in front of hundreds of people.

For sure, revealing sensitive information takes guts. But the reward of building trust by doing so has been demonstrated repeatedly in research. Yet sometimes we confuse *saying* we're trustworthy with *doing* something that actually builds trust. Assurances alone don't cut it. In fact, they can backfire.

In one study, we asked seventy-nine students a series of personal questions—some standard, some sensitive. We wanted to see how much, or how little, personal information they revealed. But before they responded, we gave them either no assurance of confidentiality, weak and informal assurance, or strong and highly specific assurance. While informal assurance sometimes encouraged disclosure, our strong and specific assurances produced the least self-disclosure. I repeat: Our *strongest* assurances made people the *least* comfortable revealing sensitive information. People were more willing to reveal when we gave them no privacy guarantees at all.

Why? Because when it comes to trust, we don't want to be told— we want to be shown. To come across as trustworthy, you have to actually *be* trustworthy. That may sound obvious, but it's surprisingly deep-rooted.

Saying "you can trust me" is easy. Too easy. It's what scholars call a "low-cost signal"—cheap to say, easy to fake. By contrast, actual vulnerability is a high-cost signal. It carries risk. And so when people volunteer something unflattering, it often makes us trust them *more*. This may also be why heavy-handed assurances can backfire. They sound like cover. Like when a loved one calls you out of the blue and starts with "I'm fine, but . . ." You assume they're anything *but* fine. Or similarly, when someone insists, "I don't lie." As Shakespeare put it: "The lady doth protest too much."

And yet we often treat self-disclosure purely as a "give." We act as if, by revealing something about ourselves, we're giving up power or leverage. Sure, that's sometimes the case. But the more I've studied disclosure decision-making, the more I've come to believe that this is the wrong frame. Increasingly, I've found myself asking: What if revealing isn't zero-sum, but mutually beneficial?

Life Is Not a Zero-Sum Game

John von Neumann believed that poker held vital clues for optimal decision-making. That's because, in his view, life is full of "little tactics of deception." "Von Neumann was only interested in poker," economist Tim Harford writes, "because he saw it as a path toward developing a mathematics of life itself. He wanted a general theory that could be applied to diplomacy, war, love, evolution or business strategy." If only human beings were simple enough to understand with such a mathematics, but we're not (thank goodness!). Not even poker as it's actually played is simple enough for such a set of rules. Von Neumann focused on a simplified two-player version, where one person's gain is precisely the other's loss, a classic zero-sum setup. And while poker is often played among multiple people, the total winnings still exactly match the total losses. The pot is merely redistributed. Meanwhile, in life generally, as games of cooperation illustrate, we need not see ourselves as being in a zero-sum competition.

Like other instructors, I make this point at the start of the first session of my negotiation course. I haul three folding chairs down to the front of the amphitheater-style classroom and put them in front of the three lucky students sitting at the desks in the front row. Then I ask for three other students to come down and sit in the folding chairs, so the six students are sitting face-to-face with the desktops between them. Now they take part in a little competition. I instruct them to arm wrestle with their partner for the next thirty seconds. Each player will get a point each time they pin their opponent's hand down, and the goal is to amass as many points as possible for oneself.

Most pairs engage in a vigorous arm-wrestling match, and each player manages to score one, maybe two points for themselves. But occasionally, a savvy pair will begin to alternate wins, allowing their opponent to easily pin their arm down right away and vice versa. This way they both rack up far more points for themselves. Another benefit is that the students who collaborate this way have a whole lot more fun.

Just as in this game, building trust requires someone to take the leap into vast uncertainty. In the arm-wrestling exercise, someone has to flirt with loserdom in front of the whole class by "giving in" to the other side, hoping that maybe, just maybe, their opponent will reciprocate. Just so, in disclosure dilemmas, being the first to open up takes guts—you relinquish control to the universe and accept the great unknown. Sure, this makes you vulnerable to bad outcomes. But it also makes you eligible for some of life's richest rewards. Those rewards are not zero-sum. They're not like dividing up a pot of money where if I get more, you get less.

When we reveal something sensitive, we are not necessarily entering a zero-sum transaction. We are creating a possibility for mutual trust, better relationships, connection, growth, even safety. Ironically, what seems like a loss of control is often what unlocks the very things we want most. This is the mistake economists (and, frankly, a lot of us) often make: treating information largely as a commodity to be protected or extracted. Disclosure is an investment—it's risk in the service of trust. Trust, unlike poker chips, isn't finite.

Yes, some things in life truly are zero-sum. Like when we're competing for a fixed amount of bonus at work, or in school when we're graded on a forced curve. Often, though, we perceive we're in a win-or-lose competition when we're not, or we don't have to be. When it comes to disclosure decisions, we'll most often be better off if we appreciate that we're not playing for a fixed kitty of chips. We can reap, and share, much richer rewards by being transparent about our needs and desires.

Transparency—revealing our goals, in this case—is central to realizing the win-win possibilities. Consider the following metaphor that I and others use to teach the point that life isn't as zero-sum as we expect it to be: Imagine two people are fighting over who will get an orange. They argue and argue for so long that the orange goes bad and neither of them gets it. But the thing is, if they had only communicated *why* they wanted the orange, they would have quickly realized that they could both have been perfectly satisfied. That's because one of them was baking a cake and only needed the rind, while the other only wanted the juice. Instead, they had each egocentrically assumed that the other shared their interests—that they both wanted the orange for the same purpose.

Looking back to my own "humiliation" moment, when I told my senior colleagues about how I peed onstage, I'm not so sure that this was the zero-sum situation that I had felt it was the day after. Yes, I had a massive disclosure hangover. But what about the long game? Several of the senior scholars in the circle that night became some of my closest friends and mentors. I wonder if my authentic confession might have actually enabled these friendships. Is it possible that I, compared to my more cautious co-confessors, benefited more from sharing something that was actually embarrassing? Am I even benefiting from telling you now? Isn't vulnerability sometimes more valuable than respectability—or success? Might it even be integral to garnering respect?

<hr />

I n the fall of 2006, I had just moved to Pittsburgh from my home and native Canada to start grad school. And as it happened, I encountered a real-world version of a game theory dilemma in my very first week of studies. I felt I had gotten into the super-competitive doctoral program against all odds, and that I had to prove myself. Or at least, avoid making an intellectual blunder, like asking a less-than-brilliant question in one of the many research presentations we attended during orientation. I perceived the entire experience as an extended intelligence test I was taking in front of my classmates.

Making matters more anxiety-provoking was that the professor who introduced the dilemma was a world-renowned game theorist.

In his presentation, he paired us up and slapped a twenty-dollar bill down in front of each pair. Then he gave us a variant of the Prisoner's Dilemma to grapple with. The game takes its name from a scenario in which two suspects are held in separate rooms, each forced to decide whether to betray the other or remain loyal, without knowing what the other will choose. In our case, this all played out in a seminar room during orientation week. The rules were simple but psychologically brutal: Each person had to privately choose whether to "split" or "keep." If both people chose "split," they'd share the twenty dollars evenly. If one chose "keep" while the other chose "split," the keeper would get the full twenty dollars and the splitter would get nothing. But if both chose "keep," then neither would get any money, and the prof would get his twenty bucks back.

I tried to read my partner. Would she be ruthless? Would *I*? Wouldn't I feel awful if I ended up with all the money and she got nothing? Wouldn't she?

When we all revealed our choices, I was the only one who'd chosen split. My partner was the only one who walked out of class with the twenty dollars. Way to go, LJ, I muttered to myself. Great first impression. You look like a total rube.

Yet, over time, as I ruminated about the episode, I realized that I maybe didn't need to feel humiliated. Because even though my new classmate hadn't made the same choice—and even though mine was "wrong" by standard economic theory (gasp!)—I learned something about myself, if only in hindsight: I'd rather take the risk of trusting and sometimes look naive than never open the door to cooperation at all.

3

Understanding Undersharing

For decades, Marilyn Mach was a genius hiding in plain sight. Born in 1946, the child of immigrants, she grew up above her German father's bar in downtown St. Louis, where he and her Italian mother worked long hours. Marilyn had a quick, analytical brain and was put into gifted classes at a young age, but her family's neighborhood was rough, and the only books in her house were her older brothers' *Hardy Boys* and an encyclopedia.

At age ten, Marilyn took the Stanford-Binet IQ test and reportedly got the highest score possible, her mental age clocking in at twenty-two years and ten months—an IQ of 228. The average American's IQ is about 98. (And yes, IQ tests are flawed, but they do still carry weight, especially when the numbers are extreme.)

Marilyn committed her score to memory, but she kept herself apart from the other smart kids in her classes, whose nerdy ways repelled her. Determined to be ordinary, she curled her jet-black hair, got married at sixteen, and graduated 178th in her high school class of 613. She raised her kids in St. Louis, invested in real estate, dabbled in writing, and divorced twice. A perfectly average life.

But Marilyn never forgot that perfect IQ score. And when her kids left home, she seemed to decide she was done being ordinary. She moved to New York City and put her keen analytical mind to work writing IQ quizzes and articles about intelligence. Somewhere along the way, she took her mother's last name, vos Savant, likely because it suggested an uncommon intelligence. She also joined Mega, a society for geniuses (she already belonged to the less selective Mensa), and in 1985 ended up in the *Guinness Book of World Records* under "Highest IQ." With this came the unofficial title "the smartest person in the world" and modest fame. She was no longer hiding.

Thanks to this attention, vos Savant started writing a weekly column, "Ask Marilyn," for *Parade* magazine. It was syndicated in Sunday newspapers across the United States, accompanied by a glam shot of her, a 1980s Snow White with voluminous dark hair and porcelain skin. Vos Savant fielded readers' logic and math puzzles and answered questions on topics from how the sun burns without oxygen to whether mixed-breed birds exist. She met Robert Jarvik, the inventor of the first permanent artificial heart—the Jarvik-7—and they fell in love. In 1989, the cover of *New York* magazine christened them "The Smartest Couple in New York." Vos Savant was finally living up to her potential.

And so it was in September 1990 that she sat down to write her weekly column. On this day, she couldn't have realized she was wading into a particularly controversial topic. She likely just wanted to write a straightforward answer to a question that a reader had raised about a classic brain teaser (now known as the Monty Hall problem, for reasons that will become clear soon). In short, the reader asked: "Suppose you're on a game show, and you're given the choice of three doors: Behind one door is a car; behind the others, goats. You pick a door, say No. 1, and the host, who knows what's behind the other doors, opens another door, say No. 3, which has a goat. He then says to you, 'Do you want to pick door No. 2?' Is it to your advantage to switch your choice?"

Marilyn's answer was simple: Yes, always switch. The odds double in your favor. To illustrate, she scaled the problem up to a million doors—making it obvious (or so she thought!) why switching works.

Marilyn turned in her breezy response and moved on to the next thing. Little did she know that her seemingly innocuous column would trigger an avalanche of adamant, angry letters insisting she was wrong—about ten thousand letters, including from mathematicians and scientists.

"You blew it, and you blew it big!" wrote Scott Smith, a mansplainer *avant la lettre* with a doctorate from the University of Florida (and maybe a tad too much coffee in his system). "Since you seem to have difficulty grasping the basic principle at work here, I'll explain. After the host reveals a goat, you now have a one-in-two chance of being correct. Whether you change your selection or not, the odds are the same. There is enough mathematical illiteracy in this country, and we don't need the world's highest IQ propagating more. Shame!"

Another PhD, Robert Smith (the Smiths of the world seemed particularly peeved, for some reason), huffily wrote in a spasm of condescension, "I am sure you will receive many letters on this topic from high school and college students. Perhaps you should keep a few addresses for help with future columns." The letters continued to rain down in this vein, and many people started to chidingly refer to the problem as "Marilyn and Her Goats."

But vos Savant was right.

You should switch. And this widespread mistake sheds light on why people make another common error: undersharing.

Let's dig deeper into this problem, which has become famous in decision science because it tells us something about how our emotions and psychology affect our decision-making—how we feel when we *do* something versus when we *don't* do something. This, in turn, relates to how we feel when we *share* versus when we *don't* share, and thus why and when we choose to share or not.

Zonked by Our Own Minds: How a Game Show Explains Undersharing

Pretend you're attending the famous game show *Let's Make a Deal*, which started airing in 1963 and was the source of the logic problem that vos Savant's reader sent in (and which I have a special fondness for from watching the German version as a kid when my family lived in Germany).

In the show, various audience members dressed in attention-grabbing "pick me!" costumes are invited to make trades with the host, the charming and perennially well-tanned Monty Hall. (Fun fact: He's Canadian, like me! Another fun fact: We Canadians like to point out who is Canadian!) Imagine Monty invites you to stand and shows you three doors onstage: Door 1, Door 2, and Door 3. Behind one of the doors is a prize: something "fabulous," Monty intones, like a "braaaaaand new car!" Behind the other two doors? *Zonks*, which are dud prizes, such as trips to nonexistent destinations, play money, broken appliances, ridiculously large amounts of food, and live animals—yes, most commonly, goats.

Monty asks you to choose a door. You choose Door 1.

Next, Monty opens one of the doors you *didn't* pick, Door 3. A dramatically ominous tune sounds out, one so ingrained in my and my brother's memory that we still intone it when something disappointing happens and we're trying to make light of it. Behind the door is a Zonk—round-trip tickets to Zonklandia. Now, before

Monty opens your chosen door, Door 1, which you hope hides a car, he wants to know: Would you like to switch to Door 2 (which remains closed)?

If you're like most people, you choose *not* to switch. You stick to your guns with Door 1. It feels right. But is this a good strategy? Will it maximize your chance of winning the prize?

The answer is a resounding no. You should always switch. If this is not intuitive to you, you are not alone. This is what led to all those angry letters to vos Savant. But the bottom line is that staying only wins on the occasions where you happened to choose the winning door from the start.

The trick is that Monty knows where the car is, and before he asks you whether you want to switch, he *always opens a goat door.* That means he's actually helping you, by taking a dud door out of the running. Think of it this way: When you first pick, there's a 1-in-3 chance you choose the car and a 2-in-3 chance you don't. So, most of the time, your first pick is wrong. Monty then shows you which of the other two doors is a goat, leaving the other door unopened. If you switch, you're betting that your first pick was wrong (which it usually is). And so, chances are, switching is a winning move. If you stay, you only win in the 1 out of 3 times you guessed right at the start. But if you switch, you win on 2 out of 3 occasions. Switching doubles your odds of winning. As Marilyn tried her best to explain to the mansplaining PhDs of the time, you should indeed switch. This is illustrated in the figure.

Marilyn vos Savant, smartest person in the world, vindicated.

Eventually, Marilyn managed to convince many of her vocal critics that they were wrong and she was right. To their credit, a few of her angry pen pals sent her apology notes. "It's been an intense professional embarrassment," one of them, mathematics professor Robert Sachs, told the *New York Times.*

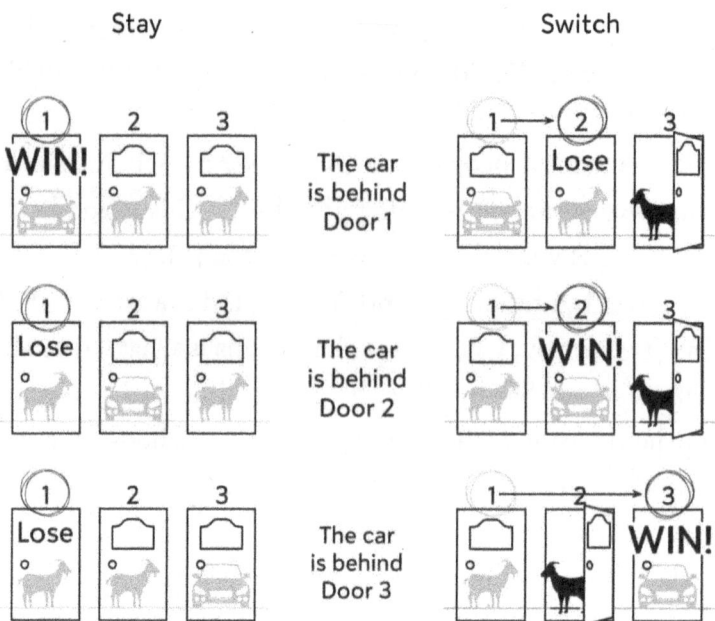

The Monty Hall problem: If you stay, you only win when the car is behind your original door (1 out of 3 chance). If you switch, you win when the car is behind either of the other doors (2 out of 3 chance).

Since then, the Monty Hall problem has been studied a lot (ad nauseam?) in behavioral science. And the thing is, even when people are told the math—or intellectually grasp that switching gives them a better chance of winning the prize—they often still resist switching. I myself might even insist on staying. But staying is illogical—you're giving up a free chance to double your odds. And yet, many of us think, "Whatever, I'm sticking with my door." Totally fair. Because there's a deeply compelling *psycho-logic* to staying.

Switching, only to discover you were right in the first place, really sucks. "If I'd only just stuck with my gut!" you chide yourself. By contrast, losing the other way—by failing to switch to a winning door (when you chose a losing door in the first place) doesn't feel nearly as bad. This is weird in a way, because in both cases, the "objective outcome" is the same: You don't win the car. And yet one hurts much more than the other. What's going on?

At the crux of this asymmetry is something that psychologists call *omission bias*, and it refers to how a bad outcome feels worse when it arises from something we *did*—an action we took—as opposed to when it arises from an inaction, or something we *didn't do*. In other words, so-called "sins of commission" (bad things that arise from our actions) are more poignant than "sins of omission" (bad things that arise because of a failure to act).

This bias makes a certain kind of sense. Actions often are more intentional than inactions. It's worse to shove someone into a fire than to fail to pull them out. And our legal systems reflect this: We generally punish harmful actions more than harmful omissions. Most states don't even require you to help someone in danger if you're just a bystander.

But omission bias can also lead us astray. We underblame ourselves for things we didn't do but probably should have done (sins of omission). And we *over*blame ourselves for things we did, in cases where our actions didn't actually cause the bad outcome. We suffer unnecessarily, convinced we did harm when we didn't—when the bad thing would have happened anyway. In disclosure dilemmas, omission bias pushes us toward silence. But sometimes, silence leaves out the very thing that matters most. And that's where the real trouble begins.

Should I Stay or Should I Disclose?

One evening, Jennifer called her mother, Donna. She urgently sought her mom's advice.

The cordless phone in her parents' house rang. Donna picked up. "Hello?"

Hearing her mother's voice, Jennifer's mind immediately wandered to the den of her childhood home. She imagined her mother starting to take a seat on the well-worn family couch. A couch where the two of them had spent many evenings cozied up, watching their favorite shows. Jennifer had always been a cuddle bug, even as a teenager. They'd start the evening sitting side by side. Then Jennifer would cozy up, resting her head on her mother's shoulder. As Jennifer often lovingly teased her, Donna didn't exactly "cuddle back." Over the course of the evening, Donna found herself subtly and slowly shifting away. Then Jennifer would cozy up again, causing Donna to slowly scooch over farther. By the end of the night, Donna invariably would end up pinned against the side of the couch. A quiet warmth swept over Jennifer as the memory surfaced, vivid and tender.

"Hello?" Donna said again.

Jennifer snapped back to attention. "It's me," she said, her voice tentative.

"What's up, my chick?" Donna said, clearly sensing something was wrong.

"I'm having second thoughts about marrying Phillip." The wedding was in just a month's time. Jennifer continued, "I love being around Phillip. He's my best friend. I just—I just—I'm not sure. Shouldn't I feel sure at this point?"

"Hmm," intoned Donna.

Then Jennifer asked: "How did you know you wanted to marry Dad?"

Donna took a deep breath. "Well, that's an interesting question. Because about a month before we were married, I had second thoughts, too. At the time, I was worried . . . you see . . . I was—I am—deeply in love with your father. But right around the time of our wedding, I couldn't help but notice how much I also really enjoyed my flirtations with other men. That made me anxious and made me question myself."

"Wow," said Jennifer. She was a little (read: a lot) caught off guard. It was rare for her mother to speak so openly about complex feelings. That was part of what made the moment so striking—and comforting.

"But I went through with it," Donna continued. "And I'm so glad I did! We've been happily married ever since. I can't imagine life without him. Were there other guys I was even more physically attracted to? Yes. But what your dad and I have is much more special. Physical attraction naturally wanes over time, anyway. I know you—you're like me: You struggle to make decisions. You doubt yourself and second-guess. I get it. I get you."

Jennifer clung to her mother's message: Attraction isn't everything; her bond with Phillip was already the "mature kind." After all, look at how happy her mother was!

Consoled by the pep talk, Jennifer married Phillip a month later. Happy ending, right? No.

From the beginning, Jennifer felt deeply dissatisfied with the marriage. She also felt shame over her dissatisfaction. She berated herself for being "shallow." *Why can't I just be happy with this wonderful, mature love like my mother?* she thought. Her eyes began to linger on other men at work conferences. She hated herself for it but couldn't seem to help it.

Five complicated years later, Jennifer and Phillip were divorced.

What went wrong? How could this story possibly fit into a chapter about undersharing when Donna shared quite a lot with her daughter? Let's rewind to that fateful phone call.

From Jennifer's perspective, and yours, too, I imagine, her mother had been quite forthcoming; she had told her daughter a very personal story. This was unusual for Donna. To Jennifer, it was refreshing and, she thought at the time, enormously helpful.

But what Donna did *not* say—what she omitted, even if inadvertently—was just as consequential, probably even *more* consequential to Jennifer's decision-making than what Donna did reveal. Had Donna not omitted some crucial, but highly personal, information, Jennifer probably would not have gone through with the wedding. She can't be sure, but she highly doubts she would have.

One morning after her divorce, as Jennifer was sipping her daily dose of coffee in her new bachelorette pad, she opened an email from her brother. He had forwarded her an email chain between their mother and father: "Mom added me to this email chain. They're talking about money plans in the most recent one, and so they added me"—her brother helped them with their finances—"but clearly they've forgotten what's at the start of this chain . . ." Dot dot dot, indeed.

Jennifer quickly scrolled through the chain, her mind scrambling to make sense of it. Finally, she pieced it together. Her jaw dropped, literally. She couldn't believe it. But then she could. Suddenly everything made sense. And when it did, her mind and body filled with anger. She thought back to that fateful phone call when her mother's words had assuaged her and given her the confidence to go ahead with the wedding. She now saw that her decision had been shaped by what felt, in hindsight, like a major omission.

What Jennifer learned was that her parents had a very different kind of marriage than she'd understood. The email chain had spelled out—in *very concrete terms*—her mother's plans to spend the weekend at a country house with a lover. The conversation, exchanged exclusively between her and Jennifer's father (well, until they inadvertently looped in Jennifer's brother—oopsies!), made it very clear they were on the same page. It was a deeply considered arrangement that worked for them (despite the occasional complications, as her parents would later admit). In some parts of the world, this might barely raise an eyebrow. But to Jennifer, it was shocking. Because it was so far from what she'd envisioned, or wanted, in a marriage. She wasn't exactly a trad wife in the making, but she craved the certainty and exclusive intimacy of monogamy with every fiber of her being.

Jennifer happily remarried a few years later to someone she was very attracted to in every possible way; her ex, too, moved on and remarried. But she couldn't help but wonder how much suffering she would have avoided, not to mention the tremendous pain she had caused her ex, had her mom not concealed this critical detail in that fateful call. Knowing only half the truth had given her false confidence. Half the story can be more harmful than no story at all.

With the passage of time, Jennifer's anger toward her mother faded to disappointment—and a lot of unanswered questions.

So she worked up the nerve, years later, to broach the topic. Jennifer's kids were in bed, and Grandma Donna was visiting. They sat down with glasses of scotch. Jennifer tried to play it cool, like this was just a casual trip down memory lane—not, you know, the culmination of a decade's worth of emotional buildup or anything like that. She said, mustering her best nonchalance, "Remember that phone call we had the month before I married Phillip?"

"Yes." Her mother smiled warmly. "I've thought about that conversation many times. And I've felt a bit bad about it. I've often wondered whether I said the right thing." They both teared up.

For the next hour and a half, time stood still. Jennifer had the deepest conversation she'd ever had with her mother. By the end of it, she felt like she understood her mother in a way she never had before, and in a way that perhaps no one else did, maybe not even Donna's husband.

As far as that fateful phone call was concerned, Donna remembered it largely how Jennifer did—but with one big, fat, and completely understandable (now that we understand omission bias!) exception: She didn't experience this as a "disclosure dilemma." The thought of sharing this deeply personal, yet deeply relevant, detail hadn't even come to Donna's mind during that conversation. Years later, when Jennifer brought it up, her mother appreciated how it would have been relevant to share. For her part, Jennifer also gained an appreciation for just how hard this would have been to share.

At the same time, Jennifer realized there had been just as much omission on her own end. Maybe her mother would have given her different advice if Jennifer had gotten up the nerve to voice her doubts sooner—and if Jennifer had been honest with *herself* sooner. A month out from the wedding, Donna probably felt pressure to reassure her daughter about a choice that already seemed like a done deal. And waiting so long to tell her mother how she felt about that fateful omission had left Jennifer stewing for years over something her mom didn't even realize she'd done (or, rather, not done).

Jennifer also recognized that part of what led her to move forward was what she wanted to believe. In the thick of prewedding anxiety, the hardest path—calling things off—was almost unthinkable. Her mother's words gave her comfort, and she clung to them. Looking back, she could see how she'd fallen into a common mental trap called confirmation bias. We tend to scrutinize the messages we don't want to hear, and let the comforting ones slide, especially when they allow us to delay, or avoid, a difficult choice.

In the end, both had stumbled at finding the right words at the right time. Like mother, like daughter. But also, better late than never. When they finally opened the correct door of sharing together, even all those years later, it felt like they'd won a prize.

The Less Said, the Better . . . or So We Think

Omission bias doesn't just explain why we hesitate to share. It also explains why, when we actually stop to think through a disclosure dilemma, we zero in on the dangers of speaking up and largely ignore the dangers of staying silent. We fixate on the *risks of revealing* and neglect the *risks of concealing.*

To test this directly, my superb collaborator Elinora Pentcheva and I ran a series of thought experiments. We asked hundreds of people to imagine dilemmas like whether to tell a new manager about their ADHD or to share a sexual fantasy with a long-term partner. In each case, we asked them to rank how strongly each of four considerations would factor into their decision:

- the **risks of revealing** (a sin of commission)
- the **benefits of revealing** (a virtue of commission)
- the **risks of concealing** (a sin of omission)
- the **benefits of concealing** (a virtue of omission)

The results were striking. Across scenarios, the *risks of revealing* almost always took the number one spot. By a landslide. The *risks of concealing* came dead last. In other words, omission bias isn't just theoretical; it reliably shapes how people think through disclosure dilemmas. We fixate on what could go wrong if we reveal while barely registering what could go wrong if we keep quiet.

If the four considerations were competing in a race, the risks of revealing would take the top spot, standing tall on the winner's podium. Meanwhile, the risks of concealing wouldn't even make the podium. They come in at a distant fourth place. The other two considerations fall in between, with the benefits of revealing coming in second place and the benefits of concealing closely behind, in third place. But the key takeaway is that the risks of revealing emerge as the most important consideration and the risks of concealing as the least important. Our structured setup, where we explicitly laid out all four considerations, makes this finding more striking. In real life, with no checklist in front of us, the risks of silence are even easier to overlook. Meanwhile, the risks of revealing scream for our attention.

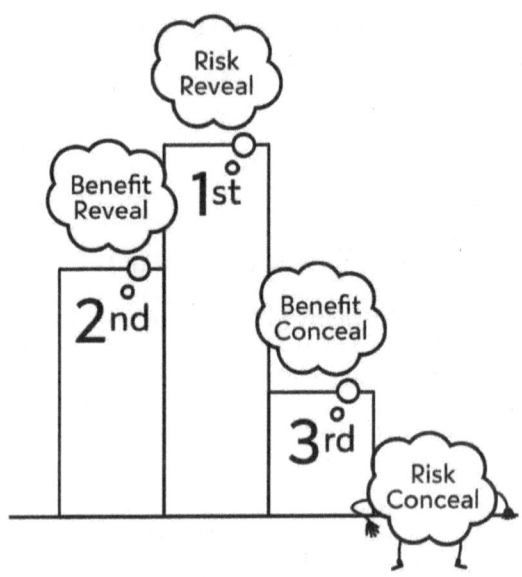

How we weigh the risks and rewards
when deciding whether to disclose

This is why sticking with silence can feel like sticking with your original Monty Hall door: the "safe" choice, the familiar one. But in both cases, that is an illusion. We're not even-handedly weighing risks versus rewards; we're reacting on autopilot. To be clear, I don't mind if you choose to stay (not that you need my permission). I think it's a perfectly defensible choice to stay if you do it eyes wide open—as in, if you reasoned: *Sure, switching doubles my odds, but that bump isn't worth the gut-punch I'd feel if I lost that way.* That's a conscious tradeoff. What I want us to avoid is the knee-jerk reflex—the instinct to stay silent simply because it feels safer (because it often isn't). The key is to see omission bias for what it is and to ask ourselves what we might be missing by holding back.

The Sting of TMI versus the Ache of TLI

Now, there's another bias that makes things even trickier: we overestimate the length of a disclosure hangover—that horrible feeling that comes after you spill a dirty secret at a party or tell off your boss. Like most hangovers, the pain of TMI is often short-lived. But we only tend to realize this when we reflect back on them. In the moment, when in the throes of TMI, it's hard to imagine it will end. And so we come to dread the mere prospect of a disclosure hangover and avoid putting ourselves in that position of potentially getting one—that is, we avoid opening up.

This is what psychologists call *impact bias*: our tendency to overestimate the strength and duration of emotional events. Win a raise? We expect the glow to last forever. Go through a breakup? We imagine devastation stretching endlessly. In reality, highs fade faster than we expect, and lows ease more quickly than we fear.

The kicker is that not only does the sting of TMI fade faster than we think, but over time it's often replaced by something heavier: the ache of not sharing. The ache of TLI.

When we overshare, the sting can be sharp. But our brains are surprisingly good at helping us recover. Imagine telling someone you love them, only to be rebuffed. There's no way to unsay the words or undo the mortification, but silver linings eventually soften the blow. You might reframe it as a blessing that you found out sooner. Or you might dig into the causes and realize it wasn't all on you. Either way, the pain tends to fade.

The ache of TLI is different. Silence is harder to reframe. In hindsight, we forget how much courage it would have taken to open up, so we end up blaming ourselves more. Psychologists Thomas Gilovich and Victoria Medvec found this pattern when they asked people to review their lives—to think about the things they regret. They found that "regrettable failures to act" outnumbered regrettable actions by nearly two to one. A full 63 percent of participants' biggest regrets came from things they didn't do. As Gilovich and Medvec put it, "We curse ourselves by asking, 'Why didn't I at least try?'" It's as if the regret compounds.

Which brings me back to our heroine, Marilyn vos Savant. For many years, she stuck with Door 1. She kept her genius to herself, played it safe, lived an ordinary life. And then, in her forties, she switched. She revealed her childhood IQ score, moved to New York, married an intellectual equal, launched a high-profile career, and became a household name. The queen of the goats lived out her own advice. We all have our Door 1s—the choices we stick with because they feel familiar, safer, easier to justify. Marilyn's story makes me wonder: How many of us are clinging to the safe door, when switching might just change everything?

I've started asking myself that question a lot more. Why? Because all these years later I still wince when I think about how long it took me to tell my first husband I wanted a divorce.

As you may have guessed, Jennifer was me.

For nearly a decade, I never told my mother, "Donna," how I felt that phone call had impacted me—how I believed the unspoken had shaped my life course. And how I'd found out, quietly, that my parents had an unconventional marriage and then tucked that knowledge away. Not just from her, but from myself.

It wasn't until I was writing this chapter that I saw the wedge for what it really was: an omission, an inaction with ripple effects. One I'd been living with for so long that I had stopped seeing it as a choice. And when I finally did talk to my mom, something shifted. That invisible wedge between us began to dissolve. The conversation I'd spent a decade avoiding—because it felt too awkward, too delicate, too likely to turn me inside out—ended up being one of the most meaningful ones I've ever had. With anyone.

Which brings me to the second Monty Hall dilemma I wrestled with: whether to tell you, dear reader, that this story was mine.

Of course I talked to my family about it. But still, I debated. Should I just leave it anonymous? Let the story stand on its own? Would revealing that I was Jennifer make the chapter too personal? Too vulnerable? Too . . . much?

This book is about how the things we leave unsaid can quietly— but significantly—shape our lives. Ultimately, I concluded that the story has more power if you know that I lived it. That I made one of the types of mistakes that I study. That I sat with the silence and stewed in the regret, not as a fancy Harvard professor, but as a daughter. It's one thing to know that omission bias is real. It's quite another to know that someone who studies it could be caught in it. For nearly ten years.

And that, I think, is the whole point.

The Monty Hall problem teaches us that the safer-seeming option—staying silent, sticking with what we know—isn't always the better one. Sometimes, revealing is the wiser move. Not because it's guaranteed to go well, but because it reflects something deeper: the willingness to step into uncertainty, to open a door and see what might be waiting on the other side.

4

Are You a Revealer
or a Concealer?

One of the first things people ask when they find out I study disclosure is, "So . . . do I overshare?"

What they really mean is, *Tell me about ME!* Which, fair—who doesn't love a little navel-gazing? People want to know if certain personality traits make them natural sharers or secret keepers, and what that says about them. People also wonder about the broader groups they belong to. Are women more open than men? Do different cultures nudge us toward revealing or concealing? And, most important, can we change our natural tendencies?

Decision-making quirks like omission bias and the ever-present fear of disclosure hangovers often prompt us to keep things to ourselves. But our unique personal characteristics also affect our disclosure decision-making and can interact with universal cognitive biases in complex ways. Let's start by looking at one of the most salient individual characteristics: gender.

The Expressivity Gap

"Boys' emotional expressions decreased with age from 4 to 6, whereas girls' expressions did not." When I first read these words a while back, my heart sank, as my eldest son was four years old. Was he already on the road to emotional suppression?

The study that shook me and made me wonder whether I was doing enough to nurture my son's emotional life was conducted by psychologist Ross Buck in 1977. In it, preschoolers looked at emotionally charged pictures while being filmed unobtrusively to see how much emotion they showed in their facial expressions. Buck also measured their stress levels. And the way he measured this? *Sweaty palms.* Yes, in the grand tradition of psychological research, Buck measured what's called galvanic skin response, changes in skin conductance caused by sweat gland activity—essentially, a proxy for internal arousal or anxiety. What really struck me was that the children who showed more emotion—whose faces gave away what they were feeling—were actually *less* stressed. They showed fewer changes in skin conductance. In other words, the kids who let it out were physiologically calmer.

That felt like a revelation. Isn't that exactly the opposite of what boys often get taught? That staying stoic, holding it in, is how you stay strong and in control? But Buck's data hint at the opposite: that hiding emotions doesn't help you calm down—it might just make you more tense.

While Buck didn't find a significant overall difference in expressiveness between girls and boys, he noticed that as boys got older— from age four to six—they became less expressive. Meanwhile, girls didn't become any less expressive with age. This jibes with what we generally see in the daily life of grown-ups: women opening up more than men. Moreover, at least some men seem to acknowledge that they could share more. In a recent study, 56 percent of male participants said that they tend to undershare, while only 35 percent of the women said that of themselves. And Buck's research suggests that the social pressure for boys to suppress emotion already kicks in by kindergarten. Yikes.

This research hit me hard. If there's one skill I hope my kids carry into adulthood, it's emotional intelligence, or EQ—the ability to understand themselves, connect with others, and navigate the messy reality of being human. Was my son already "falling behind" in emotional expressivity? I didn't know. But I was more curious than ever: Are these gender differences innate or shaped by culture and parenting?

We may never be able to pinpoint exactly how much of the gender gap in emotional expressivity is due to nature (that is, innate) versus nurture (that is, our environment). Nor, frankly, will we be able to say whether this gender gap ought to be totally closed. Rather than going down that fascinating but inconclusive rabbit hole, I'll skip ahead to how developmental psychologist Leslie Brody sums it up. She argues that both biology and social context contribute to gender differences in emotional expression, but that the environment does a lot of shaping. From infancy onward, children learn *display rules*—the unspoken norms that dictate which emotions are "appropriate" for each gender. These rules, reinforced by caregivers, peers, and culture, channel emotional development along gendered lines and amplify any biological differences. In Brody's view, nature sets the stage, but nurture directs the play.

Which is why I smiled ear to ear when my younger son, at two and a half years old, looked up at me at bedtime and said, "I love you, Mama. But sometimes I don't like you." This evidence that he's learning to articulate emotions offered a glimmer of hope that maybe—just maybe—my husband and I are doing something right in this humbling (and sleep-depriving!) adventure that is parenting. It also made me appreciate my own mother more. When I was little, she used to swap the genders of people in male-dominated professions in the books she read to me—suddenly, the doctors, engineers, and astronauts were all women. I only realized this as an adult. Parenthood is funny that way, isn't it? It makes you see your own parents with fresh eyes. And if I can do the modern-day equivalent for my sons—nurturing their EQ so that they can discover their own personal sweet spots in how much to share—I'll consider it a win.

A Conversation That Never Happened

Setting: A university café in the late 1960s. At a booth sits Erving—Goffman, that is, the esteemed sociologist we met in chapter 2. Dressed in quiet gray, sipping black coffee, the middle-aged academic juggernaut fades into the restaurant's midday bustle like a discarded black-and-white snapshot. His eyes, meanwhile, scan every diner and waiter, examining them as if they were participants in an inductive field study. He's a master observer.

Then a man of similar age, Sid, saunters over to Erving's table, wearing sandals and a grin, holding two coffees. He slides in across from Erving.

Sid: Erv, how've you been? Long time no see. Thanks for meeting up—I've been wanting to talk to you.

Erving (*nodding almost imperceptibly, expression impassive*): No problem. It seemed . . . urgent.

Sid: Yeah, it is—to me anyway! So, I know you've got people performing on the front stages and backstage of life. But I keep wondering: When do we get to drop the act?

Erving *(raising an eyebrow)*: Drop the act? What do you mean?

Sid: Like, let down our guard and be *ourselves*!

Erving *(with slight disdain)*: Ourselves? But we *are* the act. The desire to "be ourselves," as you put it, is just another performance—a culturally sanctioned one. There's no backstage without an audience in mind.

Sid *(nodding thoughtfully)*: You could be right. But I worry this leaves people lonely. If we're always managing impressions, where does intimacy come in? In my therapy practice, I've seen my patients transform—light up—just by telling the truth about themselves. And the breakthroughs! It's truly amazing, Erv.

Erving *(showing a flash of emotion)*: And *I've* seen people ruin job interviews and humiliate themselves in my classroom by oversharing their so-called truth. There's a reason we edit ourselves. It's not deception; it's a matter of social competence.

Sid: Sure, but we can get trapped behind the mask. I think people are hungry to be seen—*really* seen. Self-disclosure isn't just therapeutic; it's liberating. It's how we build trust.

Erving *(deadpan)*: Trust is nice. So are boundaries. You wouldn't want your surgeon to open with "I've been crying all week."

Sid *(with a laugh)*: Depends on the surgery.

Erving *(smirking)*: Surgery or not, some disclosures come with malpractice risk.

Sid: So, you think disclosure is risky, eh?

Erving: Let's just say I once told a date I was a sociologist of everyday deception. She asked if I was hitting on her ironically.

Sid *(grinning)*: You needed to open up. Tell her how you *felt*.

Erving: I did. I told her I felt . . . *performatively ambivalent*.

Sid *(mock groaning)*: Erving. You need therapy.

> **Erving** *(sipping his coffee)*: Therapy is just one more performance.

Okay, maybe I got a little carried away here. But I've often thought about what it would have looked like if Erving Goffman and Sid had been buddies. That's Sidney Jourard, who is not as famous as Goffman but was no less groundbreaking in his views on self-disclosure. If they *had* met, I have a feeling they would have had a lot to say about how personality shapes our willingness to open up.

Both brilliant Canadian-born thinkers of the mid-twentieth century, Goffman and Jourard, though in different fields—sociology and psychology, respectively—were both deeply interested in how we present ourselves to others. As we saw in chapter 2, Goffman emphasized self-disclosure as strategic. Jourard focused on self-disclosure as a path to psychological growth.

I don't think their ideas were just intellectual positions; they seemed to reflect different ways of being in the world. It's also interesting (and fun!) to wonder whether differences in their dispositions might jibe with their divergent views on self-disclosure. As the saying goes, "All research is *me*search."

Goffman, by many accounts, was private and guarded. He "presented himself as a detached, hard-boiled intellectual cynic, the sociologist as 1940s private eye," recalled a former student. "He had a voyeur's interest in the intimate details of others' lives, and a strong eye for the ironic and poignant." Another former student said Goffman had an "autobiographical reticence, bordering on unyielding secrecy" and "remained by choice an outsider to the sociological department at Penn."

Jourard, on the other hand, was known for his warmth, humor, and adventurous spirit. As a young man, he hitchhiked from Vancouver to California, only to be promptly arrested when he tried to get a job without a work visa. He spent ten days in jail, where he made friends with the guards and helped other detainees write love letters.

Jourard died young, at age forty-eight, yet "he lived and loved so fully in his years," said a friend in his eulogy. "Sid had many close friends; his beach house teemed with them," said another friend. "It was easy to love him." He gave public lectures barefoot, welcomed students into his life, and drove around Gainesville, where he worked at the University of Florida, in a beat-up yellow Triumph convertible with the top down.

Jourard's intellectual curiosity led him to explore everything from body image to psychedelic drugs to clinical nursing care. His fascination with self-disclosure began early, maybe even earlier than he realized. A young Sid once took some coins from his brother Harry's dresser. At dinner, Harry announced that someone had stolen his money. Sid immediately piped up, "I didn't take your fifty cents, Harry!" Oopsie-daisy. Even as a child, he seemed primed to study what gets revealed when we're trying to conceal.

But the moment that truly lit the spark for his life's work came years later, in the therapy room. "I became fascinated with self-disclosure after puzzling about the fact that patients who consulted me for therapy told me more about themselves than they had ever told another living person," Jourard wrote. "Many of them said, 'You are the first person I have ever been completely honest with.' I wondered whether there was some connection between their reluctance to be known by spouse, family, and friends and their need to consult with a professional psychotherapist. My fascination with self-disclosure led to a conceptual and empirical odyssey."

That odyssey became *The Transparent Self* (1971), a foundational text in the psychology of openness. Jourard argued that psychological health depends on our willingness to be known—not as we want to appear, but as we really are.

Which brings us to your personality and mine.

Just like Goffman and Jourard, each of us carries a unique mix of traits that shape how—and how much—we open up. Psychologists have found that while our personalities are complex, they tend to cluster around five broad, universal dimensions known as the Big Five. Together, these traits offer a lens for understanding your own disclosure style. They are:

- Open to Experience (curious, creative, unconventional) versus Closed-Minded (unimaginative, methodical, structured)

- Conscientious (dependable, self-disciplined, thorough) versus Disorganized (careless, impulsive)

- Extraverted (assertive, talkative, active) versus Introverted (reserved, shy)

- Agreeable (trusting, sympathetic, cooperative) versus Disagreeable (distrustful, aggressive)

- Neurotic (anxious, edgy, easily stressed) versus Calm (relaxed, self-confident)

Yes, some of these labels sting, but they're the real deal, tested and vetted in psychology-land.

People are pretty good at intuiting where they fall on each trait, but if you're curious to see how you score—or just enjoy a good personality test—you can find one (along with some other revealing quizzes) at revealingquiz.com.

As for myself, I am highly conscientious and agreeable; and moderately neurotic, introverted, and open to experience. That was my self-assessment, anyway—until I told my husband my results and he responded, "*Moderately* neurotic?" (Having someone who knows you well can sometimes offer a more accurate measure of your Big Five than your own assessment—if you dare!)

Now, let's make educated guesses about our two characters' personalities. Let's start with Jourard. From everything we've covered—his exuberance, his barefoot lectures, the convertible joyrides—we might reasonably posit that he was open to experience, extraverted, and agreeable. I'd also guess that he was low in neuroticism and, from his academic productivity, I'd imagine he was at least somewhat conscientious.

Then there's Goffman. Anecdotes from those who knew him suggest he was introverted, neurotic, and, let's face it, probably also disagreeable. It's harder to judge his openness, but I'd say he was probably less open-minded relative to Jourard. He was also a prolific academic and incredibly meticulous, so likely very conscientious.

Now that we've explored the Big Five personality traits and armchair-analyzed our disclosure experts, Jourard (textbook revealer?) and Goffman (textbook concealer?), can you guess which of the Big Five traits most strongly predicts an inclination toward self-disclosure? Hint: You can infer the answer from the sketches above, with Jourard embodying a revealing personality, and Goffman a more withholding one.

The most common guess is *extraversion*.

I get it. Extraverts are outgoing; they talk a lot; they're assertive. Extraversion is also the most observable of the Big Five. Others, like neuroticism, don't manifest so overtly in behavior; you have to get to know the person to detect them. But actually, extraversion is not that predictive of opening up—at least not in the way we're thinking about it in this book.

Gregariousness and comfort talking to people are not the same as comfort opening up with people. I call this the *extraversion illusion*:

We mistake being talkative for being revealing. But depth comes from disclosure, not decibels. In my research, people judge self-disclosure to be deepest when someone is willing to reveal their fears, regrets, and other hard-to-share feelings.

And yet, the kind of "expressivity" that is characteristic of extraversion is talkativeness and expression of *positive* emotions. Think about people you know who are extraverts. Do they reveal lots of *sensitive* information? Do they get vulnerable easily? Or are they "just" outgoing? They're lovely to be around because of their infectious positivity. I have a friend who is extremely extraverted, yet she struggles mightily to open up; she has lamented that she "can't be vulnerable." At first I thought this was just an elaborate humblebrag, but I've since seen how this has caused her distress. (And, recognizing her tendencies, she now dabbles in vulnerability from time to time—woot!) By contrast, I would call myself a high revealer. I tend to err more toward TMI than TLI, yet I'm pretty introverted. That said, just because my personality predisposes me in some ways to overshare, I still often struggle to share—as we saw in the last chapter when I told you how I took years to have a hard conversation with my mom about her TLI prewedding advice.

So if extraversion is not most strongly predictive of opening up, which of the Big Five *is*? Agreeableness.

Agreeableness is a bit of a misnomer, which is why people are often surprised by its link to sharing. It doesn't just mean being nice and easy to get along with. Each of the Big Five factors is made up of several subdimensions, or facets. Agreeableness encompasses compassion, respectfulness, and trust—specifically, trust in others, the belief that people generally have good, benevolent motives and that we can rely on them to be reasonable and fair.

As we saw in chapter 2, trust is essential to opening up. When we reveal sensitive personal thoughts and feelings, we don't know how others will react. We make ourselves vulnerable to being judged. Long before "vulnerability" became a buzzword, Jourard argued that trust is the foundation of authentic self-expression. "A person will permit himself to be known when he believes his audience is a man of goodwill," he noted back in 1971. "Self-disclosure follows an attitude of love and trust."

Agreeable people tend to trust others, so it follows that disclosure feels less risky to them than it does to people on the other side of that spectrum. My collaborator Elinora Pentcheva and I have done studies in which we measured people's Big Five traits alongside their disclosure attitudes and behaviors. We've found that agreeable people tend to have a lot of what we call *disclosure optimism*, or positive expectations of self-disclosure. They tend to endorse statements such as: "Self-disclosure builds strong relationships" and "Self-disclosure brings us closer to the people in our lives." Similarly, agreeable people tend to disagree with statements asserting that self-disclosure "pushes people away" and "is risky."

In chapter 3, we saw that people tend to prioritize the risks of revealing above the other considerations. But when we break this down by personality, a more nuanced picture emerges.

Highly agreeable people depart from the norm in two key ways. They weigh the risks and benefits of revealing just about equally, and they weigh the risks and benefits of concealing equally. In that sense, they approach disclosure decisions with more internal balance. (Though not completely balanced, as they still prioritize the consequences of revealing over those of concealing.) Highly disagreeable people, by contrast, are *disclosure pessimists*: They don't just replicate the chapter 3 pattern; they amplify it. They prioritize the risks of revealing way above the other three considerations, and they also very significantly de-emphasize the risks of concealing.

Now, you might be wondering: Can disclosure mindsets be

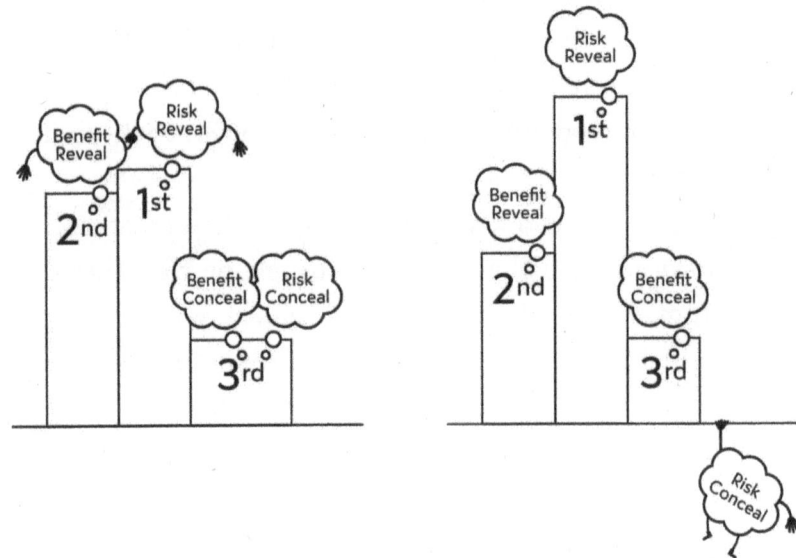

Agreeableness shapes how we weigh the risks and rewards when deciding whether to disclose

changed? Could someone with a strong tendency toward *disclosure pessimism*—someone who fixates on the risks of revealing—learn to adopt a more optimistic stance? And conversely, is it possible to be *too* disclosure-optimistic? We're still working on these questions, but one takeaway is clear: Knowing our natural tendencies can help us make better, more intentional choices. If you reflexively focus on the risks of revealing, you might want to make a conscious effort to also consider the benefits—ensuring you're not overlooking important possibilities. More on this in the next chapter.

And, if nothing else, understanding how your personality relates to your disclosure tendencies can help you to understand your foibles, as I found out. Being highly agreeable, I can be too quick to trust. Case in point: I shared news of my first pregnancy early on with our then-landlord, in a spirit of warmth and transparency.

Well, that backfired—it triggered a chain of events that ultimately forced us to move.

Toshio and Midori Yamagishi, an academic couple who have done seminal work on trust, put it this way: "A trusting person is the one who overestimates the benignity of the partner's intentions beyond the level warranted by the prudent assessment of the available information." I'd tweak this to describe people who are *overly* trusting. But the gist is, trusting others is a facet of agreeableness driving our willingness to open up.

Attachment Style and Disclosure

Attachment styles refer to the ways we form and maintain close relationships, particularly in terms of emotional closeness. They are believed to originate from early childhood experiences with our caregivers and remain relatively stable throughout our lives (though it is possible to change and become more securely attached). Of key interest here, attachment styles shape our willingness or unwillingness to open up in close relationships.

There are three basic attachment styles: secure, avoidant, and anxious. Each of us tends to be dominant in one of them. The word "dominant" is fitting because most of us have aspects of two or three styles. For example, I'd consider myself fairly securely attached, but with a dose of anxious attachment. To home in on your attachment style, try on these descriptions, adapted from psychologists Cindy Hazan and Phillip Shaver, and see which best fits:

Secure: I find it relatively easy to get close to others and am comfortable depending on them and having them depend on me. I don't worry about being abandoned or about someone getting too close to me.

Avoidant: I am somewhat uncomfortable being close to others; I find it difficult to trust them completely or to allow myself to depend on them. I am nervous when anyone gets too close, and often, [others] want me to be more vulnerable than I feel comfortable being.

Anxious: I find that others are reluctant to get as close as I would like. I often worry that [my friends don't] really love me or won't want to stay [in my life]. I want to [get very close to people], and this [sometimes scares them] away.

You might have picked up on the *anxious* side of my attachment style in chapter 3 when I mentioned my "couch antics," constantly trying to cuddle up with my mom. That was, in part, a reflection of my insecurity about closeness. I crave it, but I also worry about being rejected. I tend to seek a lot of reassurance when it comes to feelings in relationships. (TMI?)

As with the Big Five personality traits, attachment style is related to our *disclosure optimism*, or the extent to which we think that self-disclosure brings people together versus pushes them away. Securely attached people tend to be the most optimistic, avoidant people the least, and anxious folks fall somewhere in between. Attachment styles are particularly influential in interactive contexts; they show up in how responsive we are to what partners reveal.

We'll return to that shortly. For now, it's worth noting that personality traits and attachment styles don't tell the whole story. Self-disclosure isn't just an internal process; it's profoundly shaped by the external world. Where we are, who we're with, and the cultural norms we operate in all play a big role in what we choose to share. Let's start with the physical context.

Why I Travel with Lightbulbs

Social psychologists have long argued that our behavior is deeply affected by situational factors. The field itself rests on the idea that our actions are as much a function of the environment as they are of internal dispositions. After all, Stanley Milgram's obedience studies, and even Solomon Asch's conformity research, speak to the power of the situation in determining our actions. One vivid example is how the design of a space can steer us to open up or shut down.

One of the earliest scholars to explore how physical environments might shape behavior was Humphry Osmond, a psychiatrist with a penchant for the unconventional. Born in 1917 in Surrey, England, Osmond first dreamed of being an architect before pivoting to medicine. His service as a surgeon-lieutenant during World War II exposed him to the psychological toll of trauma, deepening his interest in mental health. After the war, he trained as a psychiatrist and went on to study schizophrenia. In 1951, at thirty-four, he went looking for a place to test radical ideas. He found it in an unlikely spot: a modest psychiatric hospital in Weyburn, Saskatchewan, Canada.

Here, Osmond set about rethinking the role of space in mental health. He argued that physical layout—architecture itself—could influence how people think, act, and feel. Understand the people first, he said, then design for them. Psychiatric wards, in his view, should foster connection, not isolation, and cut down on confusion. For example, vast, empty corridors typical of psychiatric wards could worsen a schizophrenic patient's distorted sense of space. Instead, smaller, clearly defined spaces could be soothing and encourage social engagement.

Osmond's work in the 1950s still shapes many of the spaces we inhabit today, for better and for worse. Look no further than Las Vegas. As documented in the design classic *Learning from Las Vegas* by Robert Venturi, Denise Scott Brown, and Steven Izenour, and further explored in Natasha Dow Schüll's *Addiction by Design*, casinos are master classes in behavioral manipulation. Casinos are engineered to quietly steer us to gamble longer. How? By stripping away clocks and windows, designing mazelike corridors with gentle curves instead of sharp corners, and carving out cozy nooks that insulate you from the chaos of the crowd. Practically every design choice keeps you immersed, disoriented, and, ideally, reaching for your wallet. Casino designers know exactly how to exploit our psychological tendencies—all the way to the slot machines.

But what does all this have to do with self-disclosure? Quite a lot, as it turns out.

Just as physical environments can be designed to assuage people with schizophrenia and goad gamblers, the spaces we inhabit can also shape how much we reveal. In a classic study, participants engaged in a simulated counseling session. Some sat in a "hard" room—bare cement floors, block walls, and harsh fluorescent lights. Others found themselves in a "soft" room, with oriental rugs, cushioned armchairs, warm lamp light, and framed pictures. The difference mattered: People opened up more in the soft room than in the hard one.

Speaking of environments that foster openness, lighting matters. That's why I like to travel with a box of lightbulbs. Not just any lightbulbs; only sub-3,000 Kelvins. A Kelvin is a measure of color temperature, from warm yellow to cool blue. In other words, coziness quantified. IMHO, warm lighting is the single easiest tweak to turn a sterile hotel room into one that feels inviting.

My office lamps are also certified sub-3,000 kelvins. I don't think it's a coincidence that I've had socially anxious academics—people who would rather analyze data than make small talk at a reception—find themselves telling me about their childhood fears.

And MBA students, usually laser-focused on impression management, have casually confessed to me their deepest insecurities as if we were longtime friends. Whether it's the glow of my lamps, my cushy rainbow-colored rug, or my pink shag pillows—all of which are decidedly *non-standard* HBS issue—I'm still struck by how people let their guard down in this environment. My office is like the Las Vegas of healthy(ish) disclosure.

Cultural Differences and Disclosure Dilemmas

In the fall of 2022, I got a firsthand reminder of just how important cultural norms can be in shaping self-disclosure. I opened my inbox to find an almost apologetic email from my literary agent (who helped me find a publisher for this book) with the results of dogged attempts to secure a Japanese publisher. My book had been pitched to nineteen different publishers in Japan, and all but one had declined (thank you, Kanki Publishing, for believing in me!). Editors noted that the book would be a hard sell in Japan. My agent's counterpart in Japan explained, "The Japanese are just more private and so don't share all the intimate details of their lives. In general, it is considered adult to be stoic in bearing one's burdens and immature to complain and burden others with your problems."

This response wasn't surprising. It fits neatly with known regional and cultural patterns.

First, there are the Big Five traits. One of the traits most different across regions is agreeableness, which strongly predicts disclosure optimism. East Asian countries, Japan included, score lower in agreeableness than any other cultural region in the world (see figure). So it's no surprise Japanese editors were wary of a book about opening up. By comparison, Africa and North America are relative bastions of agreeableness. The gaps are real, though not huge.

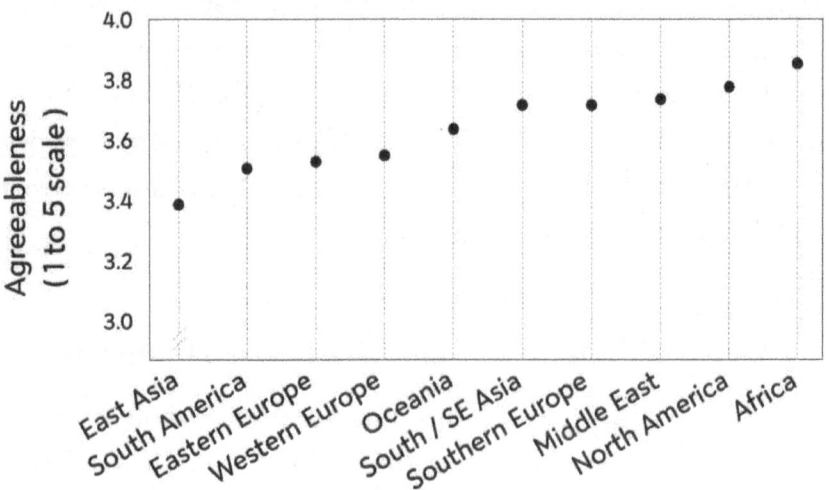

Average Agreeableness by World Region

Personality traits like agreeableness vary across cultures and help explain some differences in disclosure, but cultural values themselves matter, too. The classic lens is the individualism–collectivism spectrum, introduced by Hazel Markus and Shinobu Kitayama. Collectivistic cultures—common in East Asia and South America—value social harmony, conformity, and the needs of the group over individual self-expression. As Markus and Kitayama put it, "The emphasis is on attending to others, fitting in, and harmonious interdependence." This emphasis discourages self-disclosure, as revealing personal thoughts and feelings, especially negative ones, could disrupt group stability or make others uncomfortable.

Individualistic cultures—found most prominently in North America and Western Europe—prize independence and self-expression. Here, as Markus and Kitayama explain, people "seek to maintain their independence from others by attending to the self and by discovering and expressing their unique inner attributes." Because self-disclosure showcases individuality and forges connection, it's more encouraged.

This shapes how comfortable people feel opening up. In individualistic cultures, self-disclosure is social currency—a way to establish trust, build intimacy, and differentiate oneself. But in collectivistic cultures, disclosure is trickier. Personal sharing inherently draws attention to the individual, so it can feel antithetical to cultural values of group harmony. For this reason, people in collectivist cultures are especially cautious opening up in public.

When it comes to deep self-disclosure, cultural differences hinge on context. And one thing is for sure: Context matters even more in collectivist cultures than in individualistic ones. If you're from an individualistic culture, don't leave your situational awareness at home when visiting collectivist ones. Being thoughtful is vital to navigating disclosure dilemmas.

Stretching Your Disclosure Muscles for Maximum Flexibility

In collectivist cultures especially, context shapes what feels safe to disclose. But culture isn't the whole story; we each bring our own instincts and habits to a situation. Many of us could stand to embrace openness more than we naturally do. But some—like me, both highly agreeable and a little anxiously attached—may need the opposite: practicing restraint and seeing what happens when we hold back. The trick, in any context, is flexibility: the ability to adjust our level of disclosure based on who we're with, where we are, and what the moment demands.

Psychologists call this *disclosure flexibility*—our capacity to dial openness up or down depending on the situation. People with high disclosure flexibility can move fluidly between openness, guardedness, and everything in between. Jourard once described the hidden self as being locked behind a "psychological iron curtain," a metaphor that echoed not just the personal barriers he studied but also the geopolitical ones of his Cold War era. As we become more flexible, that iron curtain softens into fabric—something we can draw open with care rather than force.

In very close relationships, openness is vital. But in other cases—say, if you've been accused of wrongdoing or your words could harm a colleague's career—guardedness is wiser. The key is "reading the room." The easiest way to do this is to match the other person: If they're on the surface, stay light; if they're open, consider following suit. Once rapport builds, you can ease into deeper sharing. We'll talk more about this dynamic in chapter 7, which is about building friendship. A quick inner check helps, too: "What's my goal here?" As we'll see in chapter 5, asking yourself this question can refocus you on whether that goal is appropriate to the situation.

By now you won't be shocked to hear that attachment styles predict disclosure flexibility. Securely attached people are the most flexible, nimbly modulating openness depending on the situation. Avoidantly attached people have less disclosure flexibility. They tend to keep their guard up, even when a conversation partner is opening up to them. Scholars have called this pattern "compulsive closure." Anxiously attached people are rigid the other way: They are often very eager to reveal, even when it's not appropriate, as when a conversation partner isn't sharing much themselves. Anxious people are somewhat more flexible than avoidant people, but not as responsive as secure people, who are best at "matching" their partners—opening up when others do, holding back when others keep it light.

We've probably all been on the receiving end of what I call a *compulsive confessor*. *Pee-wee's Playhouse* nailed this in its 1988 Christmas special. Pee-wee's friend, famed singer Dinah Shore, calls him on his way-ahead-of-its-time magic screen. She insists on singing the entire "Twelve Days of Christmas." So Pee-wee takes matters into his own hands, hauling out a life-sized dummy of himself and plopping it down in the videophone booth while she keeps blabbing away, never noticing. Whenever a stranger overshares to me on a plane, I think of that scene. I also think of times I've done the same, wondering if the person beside me was silently wishing for a dummy of their own.

But as years of ballet training have taught me, even though people are innately different in their flexibility, one thing is certain. Everyone can get more flexible. Not just by stretching, but by strengthening. In ballet, the most breathtaking movements require both: the ability to extend beyond your comfort zone *and* the control to hold your center. The same is true of opening up. It's not about baring all, but about knowing when to let go and when to hold back.

As you move through the next chapters, I encourage you to stretch your own "disclosure muscles." Experiment. Try things on. Expand your range—not by sharing everything with everyone, but by becoming more intentional about how you open up, and with whom. You can do this while staying grounded in your natural tendencies rather than feeling boxed in by them.

Because sometimes the smallest adjustments—a softer light, a shift in posture, a few well-chosen words—can transform how open we feel, and how open others feel in return.

———○———

Sidney Jourard spent his life trying to part that "psychological iron curtain"—his own and others'. In 1974, at only forty-eight years of age, he died suddenly when the jack supporting his Triumph convertible gave way while he was repairing its starter.

One friend later wrote that he "found it hard to believe that a man so insistent upon life was now dead." He continued to say that "there was a moment, after the shock, when I was hugely angry at

Sid. Angry at that bit of arrogance, that all too typical irreverence and insouciance, that small but crucial miscalculation born out of overconfidence, which caused his demise." That anger was tethered to love—to the sense that Sid had lived wide open and was gone too soon.

And yet, in the symbolism of that convertible—top down, exposed to the wind and weather—some saw echoes of his worldview. Jourard believed that life should be lived courageously and transparently, even if that meant risking discomfort, missteps, or pain. He modeled what it meant to be a revealer. And in doing so, he invited others to step out from behind their own curtains, too.

But not everyone could. Not even those closest to him. After his sudden death, his family made the excruciating decision not to tell his mother. She had just suffered a major heart attack, and they feared that the shock of losing her son would be too much. She died not long afterward, without knowing that he had gone before her.

Was that the right call? The wrong one? Maybe neither. Maybe, like so many disclosure decisions, it was simply excruciating. But it reveals a bigger truth, one we'll explore in the next chapter: When it comes to the decision of whether, when, and how to open up, even the most openhearted among us face moments of agonizing uncertainty. And while there may be no perfect answers, there are better ways to think them through.

5

The Why of Disclosure Decisions

P aul was the baby of the family, the youngest of eleven children, born in 1965 in the small farming community of Goondiwindi, Australia. Soon after, the family moved to Toowoomba, a larger regional town, so the children could attend better secondary schools without traveling to boarding school. Paul's childhood memories are full of noise and banter—the clatter of breakfast plates, the booming voices of older siblings telling tall tales, the household's single toilet constantly in demand. But what stands out most are those special early mornings with his mother.

When he was about ten, Paul would wake up around 5:30 a.m., not out of obligation, but out of yearning for precious one-on-one time. His mother, already up, would be moving about the kitchen in her dressing gown, hair pinned up, tea towel slung over her shoulder. The house was quiet, his siblings and father still in bed. Such a big family didn't always leave much space for intimacy, but in those dawn hours, the kitchen belonged to just him and his mom. Paul would sit at the table, finishing homework, drinking warm milk, or simply listening to the tick of the clock and the soft scrape of her knife buttering toast. Sometimes they'd talk. Sometimes they wouldn't. But always, there was connection.

Looking back, Paul sees those mornings as a kind of therapy session. He asked questions—endless ones—and his mother, perhaps softening with age or just grateful for the company, answered them more candidly than she did with the others. He remembers once asking her, "Mum, why did you marry Dad? You seem like such different characters, and he drives you crazy." Her answer? "He was going off to war. I didn't think he was coming back."

For a woman of her era—born in the 1920s, stoic and emotionally contained—openly expressing her feelings was a rare occurrence. In some ways, Paul believes, their early morning rituals gave her permission to feel, to speak, and to be heard. And yet, despite the openness that flourished between them, there was one topic she would never talk about: Stephen.

Stephen was one of Paul's siblings. He had died before Paul was born, but his name lived on in the family tally. Whenever Paul said he was one of ten, his mother would correct him, "You're one of eleven." But beyond that correction, Stephen's story was locked away in a silent vault. The official narrative was brief and brutal: A toddler, unsupervised, pulled a kettle of boiling water onto himself. An accident. Nothing more to say. Except, of course, there was.

One morning, when he was a young teen, Paul worked up the nerve to broach the subject in the kitchen. His mother's response was a staccato repeat of the official version, that Stephen grabbed the kettle cord and was burned. "That's what happened. Now get ready for school." It was the only time she spoke of it. Her refusal to elaborate only deepened the mystery. Why the secrecy? Why the chill in her voice?

Soon after, Paul brought it up with his second eldest brother, whom Paul deeply admired. The brother found the story strange, too. He had no memory of the incident and had never looked into it. But he did say one thing that Paul never forgot: "If you ever find anything out, make sure to tell me."

Life went on. The siblings grew up and scattered, started families, drifted into their own rhythms. Their dad died, and their mom remarried—to the father-in-law of one of her sons. She moved to New Zealand to be with her new husband and, as Paul told me, spent the final twelve years of her life "blissfully happy." Then she, too, died. Paul, meanwhile, had become an advertising executive and eventually a podcaster. But the silence around Stephen remained—a riddle without a solution, a ghost without a grave. When Covid hit, Paul, then in his fifties, found himself increasingly drawn to the past. He started interviewing his siblings for a podcast about family secrets. And that's when Stephen's story began to resurface for him. But not with answers—just more inconsistencies.

Paul began combing through old issues of the family's local paper, the *Illawarra Mercury*, searching for any article about a scalding death in Cringila, where the family had lived at the time. And then, there it was. A short, front-page article, dated April 11, 1951—the day after Stephen had died: "Baby Fatally Scalded."

Paul almost skipped over it. His mother had said Stephen was a toddler. This was a *baby*. But there was his name—Stephen—confirming it was indeed his brother. The article gave no other details.

Paul kept digging and found another article, dated June 1. According to the piece, on April 10, their mother had briefly popped outside to the backyard to fetch something. The kettle was on the stove, and baby Stephen was in his bassinet across the room. The article recounted her testimony: one of Stephen's older siblings—a four-year-old at the time—was around, but she had called him outside after her.

The investigator concluded that between the time the mother had left the kitchen and called for her son, that son had "inadvertently pushed the basinette nearer the stove and the baby had caught hold of the cord of the kettle which had been pulled into the basinette."

Paul felt his jaw drop. "My Poirot moment," he later called it, referring to the famous detective from Agatha Christie novels. His beloved brother had accidentally caused the chain of events that led to Stephen's death. He was only four years old at the time. And so their mother had done everything possible to protect him from blame or guilt. She noted in her testimony that she had called her son outside, which may have suggested that he couldn't have been involved. She didn't lie. But she paltered, offering a technically true but potentially misleading detail. She blamed herself entirely, as any mother would.

Now Paul had a secret, like a sudden inheritance left from his mother just for him. He knew what had happened. And he remembered how his brother had told him to let him know if he ever learned anything more about Stephen's death. But should he?

Most of us will never face a disclosure dilemma quite like Paul's. But we've all felt that inner churn—that anxious tug-of-war between silence and words. Should I reveal? Should I keep this to myself? These questions haunt us in friendships, at work, in families, in love. But how do we think through them well? How do we move from gut churn to clarity?

Ask Why

It can be helpful to first take a step back and ask yourself, *Why am I thinking of revealing?* Getting clear on the "why" of an impending reveal, or the *purpose* behind it, requires a level of honesty that's sometimes harder than it looks. In daily life, we're pretty good at telling ourselves flattering stories about our motivations. We say we're protecting someone else's feelings by hiding something, when really we're just avoiding our own shame. We say we're being transparent and asking for honest feedback, when really we just want validation, or even a little admiration. So, before you can make a good disclosure decision, you have to interrogate your *why*: What am I hoping this will achieve? And: Is that really the right thing to aim for? And, if so, then what's most likely to accomplish that—revealing, or holding back?

Take my onstage peeing story, for instance. If my only goal had been to avoid embarrassment, I never would have shared that with those fancy professors. But that moment sparked real connection and helped me form lasting friendships with some of the very people I was most intimidated by—one of whom is now one of my dearest friends. In hindsight, I think I lucked out, because I certainly didn't think that decision through; I didn't understand my "why" of revealing. I think I'd say my unconscious goals in that interaction were to have fun, to connect, and, of course, to avoid eternal shame. I think I ended up revealing because in that (drunk!) moment, the fun goal prevailed. But knowing what I know now, I'd do it all over again. After all, it actually delivered on two of the three purposes, so, not too shabby. And the one goal it didn't achieve—avoiding embarrassment—was the least important and, mercifully, short-lived.

But back to Paul, who I spoke with about the discovery of his mother's secret and his process for deciding what to do next. I asked him how he thought about the disclosure dilemma. Why had he considered revealing this family secret? Unflinchingly, the first thing he said was that he values openness. Those crack-of-dawn conversations with his mother had shaped his disclosure outlook profoundly. He had learned that openness, even when it felt a little blunt, could bring people closer. As he told me, "Ever since, I've been really interested in openness, and in encouraging people to say what they mean; it's always been important to me."

So, you could say, Paul's purpose was to live a life of unusual transparency, or at least free from withholding. But he also wanted to unburden himself. Since piecing together what had really happened to little Stephen, he was now keeping a secret that weighed on him. His mind would often wander to the topic as if drawn by a magnetic force. And he'd ruminate over what to do—to reveal or not to reveal. He wanted relief.

But at the same time, he had other purposes, too. He wanted to honor his mother's legacy. He asked himself if revealing the secret she had taken to her grave would be a betrayal. After all, his mother had kept the secret for over fifty years. She had even tried to shield her son from beyond the grave. On her deathbed, one of her final, fevered whispers to him was: "Don't believe everything you read about Stephen." Paul also wasn't sure whether his brother really would want to know the truth. As Paul wryly recounted, you can't exactly ask, "Hey, if you'd accidentally killed your brother, would you want to be told?"

Paul had a good handle on *why* he wanted to reveal. But he wasn't ready to go ahead and do it. That's because, as so often happens, his different purposes didn't align. But knowing his "why" helped him frame his decision more clearly. It also inspired the almost absurd lengths he went to get a handle on the potential benefits and downsides of the decision.

So, like Paul, how can we sort through our oft-conflicting goals surrounding something we feel the urge to reveal? That's where a more structured approach can be helpful, at least for those real doozy disclosure dilemmas.

The Benjamin Franklin Upgrade

In 1772, Benjamin Franklin sent his friend Joseph Priestley a letter. Priestley, an English chemist, sought advice from Franklin. He was pondering whether to move to America or stay in England, a major life choice that felt too muddled to resolve.

In his response, Franklin empathized, writing that some decisions are "difficult chiefly because while we have them under Consideration all the Reasons *pro* and *con* are not present to the Mind at the same time; but sometimes one Set present themselves, and at other times another, the first being out of Sight." This could not be more true for disclosure decisions.

Franklin continued: "I cannot . . . advise you *what* to determine, but if you please I will tell you *how*." He told Priestley how he would "divide half a Sheet of Paper by a Line into two Columns, writing over the one *Pro*, and over the other *Con*. Then during three or four Days Consideration I put down under the different Heads short Hints of the different Motives that at different Times occur to me for or against the Measure." And, of course, Franklin aptly acknowledged that different motives have different weights.

Explicitly considering the pros, in addition to the cons, of revealing will help us; we can even jot them down, like Franklin advised his friend, like so:

Pros of revealing	Cons of revealing

Franklin's method is designed to help us slow down and think more consciously. It moves us from what psychologists call "System 1" thinking, or fast, intuitive, and emotional, to "System 2" thinking, or more deliberate, effortful, and analytical. This shift matters, especially for emotionally complex, high-stakes decisions like disclosure dilemmas. Structured reflection, even just pausing to articulate what we're weighing, can improve the quality of our decisions.

Since Franklin, psychologists have continued to use and study this type of tool. In the 1970s, Irving Janis and Leon Mann formalized it into clinical practice, dubbing it the "decisional balance sheet." They argued that the tool was particularly helpful in situations of decisional ambivalence—when we feel truly torn, when both action and inaction pull at us. Sound familiar? Yes, this is the essence of disclosure dilemmas. We're deeply conflicted.

Admittedly, I'm writing this with my tail between my legs. Because there was a time, early in my PhD program, when I rolled my eyes at anything in the "decision aid" genre—decisional balance sheets included. Back then, I saw them as toothless tools peddled by people who believed you could spreadsheet your way out of existential crises. That cynicism took root during the two loooong weeks my bestie and I spent snickering in the back row of a summer retreat put on by an Ivy League university. The purpose of the retreat was to advance the science of helping patients make more informed medical decisions using, yes, decision aids. A laudable goal, for sure. And yet there we were, two overconfident grad students armed with travel stipends, backpacks, and very little real-world perspective. To be fair, our skepticism wasn't totally off base. There had been shockingly few rigorous tests of whether these decision aids actually worked. The "interventions" in question were, in many cases, what we snidely called "glorified information pamphlets"—tools that reduced rich, agonizing trade-offs into tidy bullet points. I remember one breast cancer decision aid that listed, under *Con of surgery*, "You will lose your breast," and under *Pro*, "Greater chance of surviving." Imagine.

So yes, my back-of-the-room antics were that of a . . . well, I believe the technical term is "ungrateful shithead." The people running the show, in contrast, were thoughtful, mission-driven clinicians just trying to do their best with what they had in the messy, imperfect world of clinical reality.

The truth is, these tools can be helpful. But the details matter. The form matters. Most of all, the process matters. A decision aid isn't a silver bullet; it's scaffolding. And the key is that it needs to be active. You have to think it through yourself. Not just be handed a pamphlet with a table of preprinted pros and cons.

I see their potential utility, especially for a more modest, but no less meaningful, goal: thinking things through. Surfacing your ambivalence so you can finally examine it. That was the original intent. As psychologist William Miller put it, "The aim is not to challenge defenses but to evoke them into conscious awareness for examination."

That's exactly what we're after here.

Disclosure dilemmas are rarely simple, but Franklin's method gives us a starting point. And it becomes even more powerful when we adapt it to the specific challenges of revealing personal information. Instead of a single list of pros and cons for disclosure dilemmas, it's particularly helpful to distinguish between the consequences of revealing and the consequences of not revealing. That's because, as we saw in chapter 3, we tend to focus on what happens when we act and neglect what happens when we don't. In other words, the bottom row of this matrix—the rewards, and especially the risks, of concealing—often goes unexamined unless we bring it into conscious view. Like so:

	Pro	Con
Reveal	- - -	- - -
Do Not Reveal	- - -	- - -

This four-quadrant grid helps uncover hidden stakes, especially the oft-overlooked costs of staying silent. Once we've clarified our why, or what we're hoping to achieve, this framework helps us evaluate which course of action—revealing or not revealing—is most likely to deliver on that goal. It forces us to consider the distinct consequences of both action and inaction—including the hidden emotional, cognitive, and relational tolls of withholding, which, as we saw in chapter 3, we're especially prone to ignore. That's why the traditional pro/con list needed an upgrade.

The Truth About Secrets

Psychologists have found that keeping a secret isn't "inert." It takes active mental work. And not just for secrets, but for any piece of personal information we're actively weighing whether to reveal. In other words, any disclosure dilemma.

In a landmark study, Daniel Wegner and colleagues showed that when you try not to think about something—famously, a white bear—it can cause you to become preoccupied with that very thing. This is an example of the ironic process of mental control. When we try to suppress a thought, we can actually make it more likely to intrude. And the same goes for secrets. In a follow-up study, people asked to keep a secret performed worse on a Stroop task, a cognitively demanding test that requires naming the color that a word is printed in while ignoring the word itself (e.g., saying "black" when the word "white" is written in black ink). Those keeping a secret did worse on this task, suggesting the secret was intruding on their attention. The more we try to suppress a thought, the more likely it is to stick.

Meanwhile, other research speaks to the wide-ranging tolls of concealment. For example, the more frequently people think about their secrets, the more likely they are to experience feelings of inauthenticity, lower relationship satisfaction, and even worse physical health outcomes. The active process of managing concealment—of filtering speech, rehearsing responses, scanning situations for risk—can deplete cognitive resources, impairing focus, memory, and self-control. As James Joyce put it, "Secrets, silent, stony sit in the dark palaces of both our hearts: secrets weary of their tyranny: tyrants willing to be dethroned."

This was certainly true for Paul. The longer he kept the secret, the more it occupied his thoughts. He felt the weight of it daily—not just because it was a painful truth, but because it was a mystery he had solved, yet one that others, especially his brother, still lived with. The asymmetry was glaring. But often, our secrets and disclosure dilemmas aren't this obvious. We suppress feelings without realizing it. We keep things to ourselves and convince ourselves it's no big deal. We choose not to share, and it feels like inaction—but as we saw in chapter 3, omission is still a choice. And it can carry more weight than we expect.

Moreover, withholding a secret is a decision you can revisit. That's part of its appeal. You can always choose to reveal it later. This optionality can feel empowering. Who doesn't like to keep their options open? But research shows that having that option doesn't always serve us well.

Psychologists Daniel Gilbert and Jane Ebert found that people who make irreversible decisions, for example, choosing a piece of art they can't return, actually end up happier with their choice than people who can change their minds. When we know a decision is final, we commit. We adapt. But when a decision is reversible, we stay in a state of cognitive and emotional limbo. This pattern holds in high-stakes real life, too. In one striking study, researchers found that colostomy patients who knew their condition was permanent reported higher quality of life than those who believed it was temporary. The possibility of reversal, it turns out, can interfere with adaptation.

This has real implications for disclosure dilemmas. A decision to reveal is irreversible. You can't unring the bell. And yes, that's daunting. But withholding, while seemingly safer, keeps the decision open—perpetually open. It creates an illusion of control while potentially inviting a cycle of rumination, reappraisal, and delay. The ability to postpone can keep us stuck.

That doesn't mean you should rush to disclose. But it does mean that, as you weigh your options, it's worth asking not just what you're choosing, but what kind of decision you're making. Is it one you'll have to revisit again and again? Or one that allows you to move forward?

That's one of the virtues of the decisional balance tool. It forces us to consider not only the potential benefits and risks of revealing, but also the very real—and often invisible—costs of staying silent.

Notably, those costs are only one quadrant of the decision. In Paul's case, he'd also considered the others: the benefits of revealing, the drawbacks of revealing, and the benefits of concealing. These mapped closely to his underlying motivations. He valued openness, and revealing would have aligned with that principle; he told me he wanted to be honest, to live with integrity, to be true to himself and his beliefs. But it wasn't just about unburdening. He believed in truth-telling as an act of care: "There's no shame in any of this. We've got to talk about things," he told me.

But he also cared deeply about his mother's legacy. At first, the prospect of revealing seemed like a nonstarter. "Holy shit," he remembered thinking. "Mum kept that a secret for fifty years. Am I comfortable betraying her wishes?" It took time and thought to reconcile that. "Mum was just of an era," he told me. "People were very private. Of course she wouldn't want that fact to get out—she wouldn't want her son to have that burden. So it's completely natural that she kept it to herself." But then he followed this train of thought further. "If she'd still been alive when I found this out, I would've gone to her first and said, 'I know what happened. I know you're intensely private, but we've got to talk about this.'" He believed she would have understood. "So yes, I decided if she had still been alive, I'd have revealed it. I actually felt good about that." And, he added, "Dad wouldn't have cared either way."

Still, there were real risks to revealing—chief among them, the risk of harming his brother. What if his brother couldn't live with the news? That uncertainty loomed large. But if his brother truly wanted to know, then maybe certainty would bring closure, maybe even greater closeness as siblings. That's what made his brother such a wild card. His response had the potential to redefine the equation entirely.

Simulating the Reveal:
The Weekend That Changed Everything

This brings us to a deeper challenge with disclosure decisions more generally. Even if we've mapped out the pros and cons, it's still incredibly hard to predict how others will feel—or how we'll feel—once the decision is made.

As we saw in chapter 3, people tend to struggle with predicting future emotional reactions. We routinely overestimate how good (or bad) things will feel and how long those feelings will last. And the more unfamiliar, emotionally loaded, or high-stakes the situation, the worse our predictions tend to get. Paul's dilemma hit every one of those trip wires. So he tried something researchers recommend: Simulate the moment as vividly as possible, and pay attention to what unfolds.

Like the producer of a reality show, Paul tried to create a situation that mimicked the moment of revelation—not to actually reveal the secret, but to assess whether his brother would want to know the truth. That single question had enormous bearing on both the risks and potential benefits of revealing. If his brother showed signs that he *wanted* to know, the prognosis for disclosure looked far better. But this was one of those maddeningly circular situations: You can't really know how someone will feel about knowing . . . until they know.

Paul remembered his brother's throwaway line from years earlier: "If you ever find anything out, make sure to tell me." But he couldn't take that at face value.

So, Paul said, "I came up with this whole ridiculous construct of having the family reunion."

The idea was to bring all ten living siblings together for a weekend—a rare feat in such a large family. Paul pitched it to his siblings as a kind of living tribute. They were all getting up in age and hadn't gathered in a long time, he wrote in his invitation. Being one of eleven children is something special, he said. "Before everyone dies, we should just get together," he wrote with his characteristic openness. It was sincere. Paul really did want to get together as siblings again. But, secretly, he also hoped he'd get some resolution on his disclosure dilemma. Nearly everyone replied right away: They were in.

Paul rented a big Airbnb in Goondiwindi, where they had once lived. He planned every detail, setting an "itinerary designed to stoke memories," he said, and even creating a family trivia quiz game to provoke conversation. He passed around questions he'd written about childhood adventures, like who fell off the motorbike in 1974? He intentionally loaded the quiz with 1950s-era questions, hoping to prompt someone to mention Stephen.

Planning complete, the weekend came—and with it, Paul's chance to observe. As the siblings arrived and the conversations began, he watched carefully for any natural opening, any flicker of curiosity that might indicate that someone, especially his second-eldest brother, wanted to know more. Stephen's name came up—of course it did—but there was no sign that Paul's brother was seeking answers. No hint of lingering questions. Just fond, untroubled reminiscence.

The weekend rolled on. They visited the racetrack where their dad used to take them, attended Sunday morning mass at St. Mary's Church, toured their old school, and stopped by their childhood home. They even planted a makeshift plaque at the local pool, jokingly renaming it after their father, who, according to Paul, "had conducted illegal gambling nights to raise funds for its construction in the 1960s." Nostalgia ran thick. "We were all soon reveling in the past," Paul recalled, "our present-day selves all but forgotten. We were just kids once again, gathered around the family table."

But still, no talk of how Stephen had died.

By Sunday night, Paul was starting to panic. "I'd done everything I could to create the environment where a conversation about Stephen would come up," he told me. "But nothing. I got lost in the pure joy of the weekend." The clock was ticking.

The trivia game that night was his last chance. The night was winding down. The cabin was cozy, the wineglasses were empty, and the laughter echoed. Trivia cards were scattered across the table. Paul realized it was now or never. So he went for it, asking his brother: "Would you want to know if you had been involved in Stephen's death?" It came out surprisingly naturally, as part of the ongoing conversation, which had been filled with remembrance. His brother responded with something to the effect of "I've always wondered what really happened to Stephen."

Paul's brother paused. Then, as if on cue, he continued. He said he'd always suspected that he might have had something to do with it. Or maybe he and their sister, who had also been present that day, playing with him. It was the first time he had acknowledged these hunches aloud.

It was like a long-locked door had creaked open. The moment Paul had been trying to midwife into existence—carefully, delicately, without forcing it—finally happened.

Paul had his answer. Was this the moment to finally tell his brother?

Practice Makes Perceptive

Disclosure decisions can be *com-pli-cated*. However, I'm not suggesting we apply the Ben Franklin treatment to every little decision (though if you've got the time and interest and are super into charts—hey, why not?). After all, many, possibly most, disclosure decisions don't even rise into our awareness as such. We don't need to go to the lengths that Paul did to try to predict how his brother might respond to the news. But the decision tool can help us get a handle on the doozies—the fraught disclosure dilemmas that make your stomach twist a little. If we don't engage in a careful thought process about pros and cons, we might default to silence without even considering the alternative. Omission bias and the comfort of inaction make "keeping it to ourselves" feel like the natural choice. My hope is that, at least at first, slowing down and making these decisions more consciously—more thoughtfully, more deliberately, with eyes wide open—will help train your mind to start doing it more automatically.

And so although I recommend the tool for those doozy disclosure dilemmas, for what it's worth, I've found that using the disclosure-adapted Ben Franklin tool even for lower-stakes decisions can be a great way to practice thinking things through.

I used it recently in an area I often struggle with: telling people how I feel, especially when I expect them to be disappointed or when my feelings seem at odds with theirs (I've mentioned I'm an agreeable, people-pleasing Canadian, right?). So I typically don't share them (and when I do, I sometimes burst out in tears—yes, I'm working on this).

I faced a simple disclosure dilemma with some coauthors. We were finalizing the paperwork for a research project, including documenting who contributed to which parts of the work. This involved naming who played a leading role in categories like "idea generation," "study design," "data analysis," and "writing."

I adore my coauthors; over the years, I've been lucky to work with people who are smart, fun, creative, generous, and kindhearted (I stick to a strict "no jerks" rule when deciding whom to collaborate with, and I'm so fortunate that my job allows me that discretion). But for this particular project, when I saw the draft contribution form, I felt a small but sharp pang. For the "idea generation" category, the draft indicated we had contributed equally. I appreciated how this equal allocation might seem fair. After all, our ideas had evolved so much through collaboration that ownership felt communal, which is a lovely thing. Still, I knew the original seed of the idea had come from me—because, like so many of my research ideas, it traced back to my childhood.

Growing up around delightfully quirky decision-makers has sparked more than a few of my ideas. There were the family ski trips, where my parents' mission was to ski so many runs that we got the "price per run" down to two dollars—even though the lift ticket was already a flat fee. There was the time my mother returned from the grocery store with ten cans of coconut milk, explaining that there had been a limit of ten—which, ironically, made her feel obligated to buy the maximum. And on our family camping trips, there were the endless quests to find a "perfect" tent site—inspecting the shade, the slope, and the softness of the ground, which led us to move again and again, even when the new spot wasn't any better than the first. There was even the way my childhood babysitter assumed that just because I'm female my favorite color must be pink, which made me so stubbornly determined to favor blue that it remains my favorite color to this day. Looking back, I'm not sure I ever stood a chance of *not* becoming a decision scientist!

In any case, I remembered having the idea, and I remembered inviting my peeps to join me because I knew I couldn't execute it alone, and it'd be fun! It irked me not to see this reflected in the draft sheet, but I didn't even consider saying something. It felt too petty, too self-interested. Still, the feeling lingered. It gnawed at me. When I mentioned it to my therapist, he gently encouraged me to step back and think it through.

At that point, I paused to think about my purpose. Why would I want to say something? Not just what I *felt*, but what I was actually hoping to achieve. My why was twofold: I wanted to feel seen—credited for the idea I had originated—and I wanted to preserve my kindred friendships with the coauthors. And like any good therapist would, mine nudged me to go even deeper: Why was credit so important to me? Was it about fairness? Identity? Ego? (Therapy, if you haven't tried it, has this slightly annoying habit of making you question everything—though, mostly for the better.) Even if my purpose had layers, knowing the top-level ones—recognition and connection—helped me clarify what I was aiming for. And once I knew that, the 2 × 2 matrix became more focused.

I reeled off the downsides of revealing very quickly. "They will think I'm petty." "The lead coauthor might feel bad." In addition to, "I will burst into tears, which will be embarrassing." The upsides of holding back quickly followed. "I will avoid an uncomfortable interaction" and maybe "maintain relationship status quo."

	Pro	Con
Reveal	- acknowledgment for my ideas - authentic & true to myself - strengthen relationships?	- undermine lead author? - hard! crying (eek) - jeopardize relationships?
Do Not Reveal	- dodge the awkwardness - relationship maintenance	- rumination - feeling undervalued - letting myself down

I couldn't think of any other more compelling upsides of holding back. Which was telling.

Meanwhile, the benefits of revealing began to emerge. At first, they were simply the polar opposites of the downsides, like "They might not feel bad." But then bona fide benefits began to emerge, like "I will no longer ruminate about whether to speak up" and "If I open up, they might respect my courage." And even "If I open up, they will learn more about what I value." I thought about it some more before adding "We might even become closer."

I stared at the one empty quadrant until I could no longer bear its increasingly conspicuous absence of text. Finally, the downsides of holding back asserted themselves. "I will keep ruminating about this" and "I will feel undervalued." And then (and this one, I'll admit, required some therapist coaching), "I will be disrespecting myself by not standing up for myself."

Suddenly the downsides of revealing didn't seem so scary. For example, "What if they think I'm petty?" Well, they might. But they would also learn more about me—that I care about ideas and honoring where they come from. That getting it right matters. These are good things.

More important, they will see me for who I am. They will understand me better. And from understanding—the kind gleaned only by opening up, by getting vulnerable—comes closeness. I'd gone from not even considering opening up to having it rise to the top of my list.

If my purpose had been different, the whole decision might have tilted another way. Say, for example, that my main purpose was to keep the peace at all costs, to avoid any tension or disruption on a team that was otherwise working beautifully. Or if I had been in the middle of a stressful semester, I might not have had the emotional bandwidth for a potentially uncomfortable conversation, no matter how principled it might be. In either case, the 2 × 2 table would've looked different. The costs of revealing would've weighed more heavily, and the benefits of concealing—like maintaining harmony or sparing myself short-term stress—might have taken priority. That's the thing about this tool. It doesn't give you *the* answer, it helps you to figure out *your* answer.

Sometimes when you slow down and think it through, you realize that *not* revealing is the better choice. That was true for Paul's mother. For her, Paul reckoned, the burden of withholding was well worth its benefit: protecting her living son. "It was no wonder she concealed these details to her death in 2006, age eighty-two," Paul said. "Silence could never bring Stephen back, but it could protect the other innocent victim." Looking back, Paul respects their parents' decision all the more. As he recalled in his conversation with me, with tears welling in his eyes, "All those years when he and my older brother would be having fights, Dad never said it. He never, ever let on."

Moments We Tend to Misjudge

People often ask me, Are there certain kinds of disclosure dilemmas that we're especially prone to mishandle on impulse? In addition to the big, fraught doozies, there are other moments where the Ben Franklin tool can be just as helpful—because they're the kinds of situations where our instincts can quietly mislead us.

One such case? Gossip. That moment when someone's name comes up and you know something spicy, maybe a little shady, maybe just embarrassing, and you feel the surge: *Say it!* The urge to spill is undeniable. It's bonding, it's fun, it's easy. But if we slow down long enough to weigh the longer-term costs and benefits, we often find that the cons pile up quickly. Gossip may create a fleeting sense of

closeness, but it also makes the listener question whether you're someone they can trust. In a phenomenon called "spontaneous trait transference," people tend to associate the traits you describe with
you—even if you're talking about someone else. Say enough things
about your mean coworker, and people may start to see *you* as petty
or mean-spirited, even if they don't consciously realize why.

On the flip side, there are kinds of sharing we *underdo*. For instance, "positive gossip." Compliments. Admiration. Respect. Telling
someone, say, an acquaintance or a colleague, that you think they're
great at what they do, or that you really enjoyed something they
said, can feel awkward. But when you actually do it, it's almost always well received. Research finds that people consistently underestimate how much others appreciate these moments of warmth. We
fear being "too much." But more often, we're not quite enough.

Then there are the little disclosures about boundaries or needs—
the things that aren't emergencies, but still matter, like my dilemma
about whether to share my concern with my coauthors. These are
often the hardest to voice because they feel "minor." And because
they feel minor, we tell ourselves they're not worth mentioning. But
these little things can accumulate and harden into private resent-
ments. If I hadn't told my coauthors that I thought I should have
credit for my idea, for example, I might have become aggrieved to
the point of avoiding working with them in the future.

So no, not every moment demands a full four-quadrant reckoning. But if a decision feels tricky, emotional, or oddly lopsided—if
you're feeling pulled, or reactive, or confused—the Ben Franklin
tool (or your inner approximation of it) can help you step back, slow
down, and see the stuff your gut might be missing. Sometimes you'll
still decide not to say anything. But you'll know why. And sometimes, like Paul, or like me, you'll decide to speak up—and find that
what you feared wasn't nearly as bad as what you'd imagined.

⸻ ◇ ⸻

I n the end, Paul didn't tell his brother about the family secret that night; the moment didn't feel right. But a couple of weeks later, he called him.

"You know how you said to tell you if I ever found anything out?" Paul said to his brother. "Well, I did. And the reunion—that's actually why I organized it. I just needed to know if you really wanted to know."

Paul then told him how Stephen had died.

His brother took it in. Then replied, simply, "Thanks for letting me know. Good to know. Doesn't change anything." The implication: He couldn't blame himself if he'd made a completely accidental mistake, no matter how terrible the result. After all, he had only been four years old at the time.

It was, almost annoyingly, the exact response Paul had been bracing for—and also hoping for. A quiet, steady confirmation that yes, his brother had wanted to know. And yes, he could live with it. While people often misjudge how they'll react to difficult truths, Paul's careful effort to simulate the moment had worked—not just for him, but maybe even for his brother, who'd had a chance to get concrete, too. And for Paul, that brought a flood of pride and relief. The weekend, the scheming, the risk—it hadn't just felt worth it. It had helped him get it right.

Paul's brother passed the phone to his wife. "Tell her!" he said. His brother's wife got on the line and started to cry. Not out of sadness, but relief. Closure. "That's an incredible story, Paul," she told him. "Thank you. Thank you for telling us."

And so, the secret that had lived in silence for seventy years had finally come into the light.

In Paul's case, it wasn't just his brother or his brother's wife who found peace. A few days after Paul shared the story with his eldest sister—the one who had been playing nearby when the accident happened—Paul got a phone call. His sister was in the hospital, nearing the end of her life. And she told him something that floored him.

"After you left," she said, "I had this blister on my leg from my kidney failure. It burst. Hot fluid went everywhere. And the first thing I thought of was Stephen. It was just like the boiling. But you know what? It was cathartic."

Paul described it to me with a mixture of awe and disbelief. "It gave her this incredible sense of closure. I know it sounds bonkers, but it really meant something to her." His sister died just a few days later.

And that wasn't the only transformation the story set in motion.

What began as a carefully orchestrated plan to test whether his brother wanted to know the truth ended up becoming something much more. Paul told me that his siblings were "immensely grateful about the weekend. One of my brothers still says it was the best weekend of his life." Another sibling, a sister who had become reclusive in recent years, reemerged. "The others catch up with her a lot now," Paul said. "She's much more connected."

That weekend, Paul had tried to simulate a conversation about Stephen. But Stephen was already there. In the stories, in the memories, in the laughter that stretched late into the night. In a sense, the act of opening up brought him back. And that's an often-overlooked benefit of revealing. Sometimes, when we dare to say the quiet thing out loud, we revive something—or someone—that's been lost.

The Three Things About Doozies

As you begin to use the Ben Franklin tool, I suspect you'll notice a few things.

Thing 1: You might start seeing disclosure dilemmas every-where. (This has certainly been my experience while thinking and writing about them.) This is great news. It means you're becoming more attuned to the real prevalence of these decisions. As you read this book, you might find that at first, disclosure dilemmas cause you more—not less—mental turmoil. But fear not. You're simply becoming aware of something that's always been there. And that awareness is the first step in tackling disclosure dilemmas with clarity and confidence. Also note, the point of this tool isn't to disclose more, per se. It's to become more conscious of these moments in the first place—and then to practice thinking through them more thoroughly, like Paul did. To move, gradually, from default silence to deliberate choice. And from gut churn to clarity.

Thing 2: There's rarely a single "right" answer. The goal of the tool isn't to prescribe a path; it's to help you see the terrain more clearly and with more nuance. It helps you notice what you might be missing, especially the tendency to underplay the risks of silence. Then it helps you weigh your options against your purpose and values. And here's a hidden bonus: Simply going through this process can make you feel better about whatever you choose. Research on procedural justice shows that when people feel a decision was reached through a fair, thoughtful process, they're more likely to accept even disappointing outcomes. If you've done due diligence beforehand, at least you can rest easy knowing you cared enough to put in the effort. In that sense, this whole exercise might even serve a second function: reducing postdecisional regret. And if that's just a fancy term for rationalization . . . well, so be it. If it helps you sleep at night, I call that a win.

Let's call that Thing 3: This process may help you feel more at peace with your decision, whatever it ends up being.

In the chapters that follow, we'll explore how to harness the power of self-disclosure in service of some of life's most important ends: well-being, friendship, love, thriving at work, and courageous leadership.

SOME QUESTIONS TO ASK WHEN WORKING THROUGH DISCLOSURE DILEMMAS

	Benefits	Downsides
Reveal	What are the possible benefits of revealing? Might sharing strengthen trust, as opposed to threatening it?	What are the possible downsides of revealing? How important are these downsides? Will they be lasting or short-lived?
Do Not Reveal	What are the possible benefits of holding back? How important are these benefits? Will they be lasting or short-lived? Would I be withholding because it's the best choice or because it feels easiest?	What are the possible downsides of holding back? Will keeping it mum be burdensome or painful? Will I ruminate? Ten years from now, will I regret having held back?

6

The Healing Power of Revealing

O n the morning of Monday, May 10, 1943, in the midst of World War II, a telegram was generated at the Royal Canadian Air Force office in North Bay, Ontario. Like any other day during the war, the office would have been busy handling communications, dispatches, and updates from overseas. On this morning, the telegraph machine whirred to life with a message from the Casualties Officer. First, the address:

TO: MR M A DUQUETTE 323 MCINTYRE STREET EAST
NORTH BAY ONTARIO

Then the message itself:

REGRET TO ADVISE THAT YOUR SON PILOT OFFICER FREDERICK
ALBERT DUQUETTE J17243 (FORMER NUMBER R90502) IS
REPORTED MISSING AFTER FLYING OPERATIONS OVERSEAS
MAY EIGHT STOP LETTER FOLLOWS.

The numbers—the officer and former enlisted service numbers—were meant to ensure accuracy. But there was no mistaking the core message. Frederick Albert Duquette was missing.

The telegram would have been sealed in one of the standard yellow envelopes that had become grimly familiar across wartime Canada. It was likely delivered on foot to the Duquettes' family home on McIntyre Street.

But no one was home to receive it. Moses and Nellie Duquette were out of town, visiting Nellie's brother—a trip memorable only in retrospect as the last day before everything changed.

The Duquettes were a well-known family in North Bay. And not least because they had six sons, five of whom were serving in the war. There was Cecil, the eldest, steady and rooted. Elmer, the second-eldest, was serving overseas in Europe. Next was Bert, who along with his bride Patricia, was stationed in Centralia, Ontario, as an RCAF pilot trainer. As a new father, he was relieved of active duty. Art was also abroad, flying with the Royal Air Force, often alongside his brother Fred in Malta. Fred had always been something special. A gifted pilot stationed in Malta with the RAF's 272 Squadron, he had recently been promoted and was beloved by his squadron. Everyone called him Duke. George, the youngest, had just arrived in Halifax, preparing for deployment with the RCAF.

Nellie and Moses were my great-grandparents. Fred was their son, my great-uncle. Bert and Patricia were my grandparents. I never met Nellie and Moses, but I've always felt a connection to them, especially to Nellie. Maybe it's admiration. She raised six boys, triple the number I'm raising, and she did it during a war. Or maybe it's a sense that we shared a similar temperament. From what I've gathered through letters and family stories, Nellie seemed to bring a quiet joy to everyday life. My aunts have told me it was the kind of joy that made you feel she genuinely wanted to share it with you. She was a bit of a character, too. She had a habit of cheating at cards, with an endearing smirk that gave her away.

It was no more than a day, maybe two, before the telegram caught up with Nellie and Moses. And whatever hope they may have clung to was extinguished soon thereafter, when a letter arrived by boat from Malta, where Fred had been stationed. And I know all of this because of the careful work of my second cousin Jamie Duquette, a Canada Post worker turned family historian.

The letter confirmed what my great-grandparents had feared. "It is with deepest regret that I have to inform you of the loss of your son on the 8th of May." It went on to explain that "whilst flying between Egypt and Malta he encountered an unusually fierce headwind and ran short of petrol some 20 miles from land. He and his Observer took to their parachutes, wisely rather than risk crashing in the heavy sea, but—although a thorough and extensive search was immediately carried out by several aircraft and a rescue launch, they could not be found." It concluded by noting, "Your son was respected and admired by us all, and acknowledged to be one of the finest pilots in the squadron."

So there it was. Fred was gone. His Beaufighter had run out of fuel over the Mediterranean. He and his navigator, Sgt. Albert Johnson, had bailed out near Malta. A search team spotted a deflated dinghy and an oil slick, but no bodies. For days, aircraft and rescue launches combed the sea. Eventually, the search was called off. Fred and Albert were declared presumed dead.

An inventory of Fred's personal effects arrived at the Duquettes' doorstep sometime later. Exactly twenty items: one signet ring, three handkerchiefs, one shaving brush, one watch strap, one Bible, one cigarette lighter, one leather wallet, two photographs, one small mirror, one RCAF-issued comb, one metal identification disc, one set of dog tags, one pair of gloves, one fountain pen, one pair of cuff links, one leather-bound address book, and one blue sock.

When I learned of this family history, my mind immediately went to Nellie. I wondered how she got through it. Did she reread

Fred's letters? Did she clutch that lone blue sock tightly in her hands? How do you cope with this kind of loss? I don't know. But I do know that she didn't face it alone.

According to family stories passed down, when she and Moses learned the news, their dear friend Debney came over and stayed for hours. He had recently returned from service himself. I imagine the three of them gathered in the kitchen, warmed by the coal stove, the kettle refilled again and again. Maybe they shared stories of Fred— his flying escapades, his humor, how he would buzz low over his sweetheart's house in Malta when returning from missions. Nellie also had a special closeness with Bert, my grandfather. They wrote often and warmly, especially in the wake of Fred's death. Oral history and surviving letters tell me that Nellie didn't keep her grief bottled up. She talked about it and wrote about it.

Around the same time as I was learning about Nellie's experiences, I was immersed in the work on therapeutic effects of self-disclosure—how writing and talking about painful experiences can help us. That work began in earnest in the 1980s with psychologist James Pennebaker's landmark studies on expressive writing. Since then, a wealth of research has explored the power of talking as well as writing—from affect labeling and emotional reappraisal to therapy and even the unexpected joy of revealing ourselves to others.

In Pennebaker's studies, he asked people to spend just a few minutes writing honestly about something painful—no grammar rules, just raw thoughts and feelings. Across dozens of studies, this simple exercise modestly but reliably boosted health and mood. HIV-positive patients randomized to Pennebaker's writing task showed better T-cell counts than those given a placebo task. University students who wrote about traumatic events made fewer trips to the health center. Recently unemployed adults found jobs faster. For such a light intervention, the payoff is surprisingly big.

The one factor that consistently moderates the benefit? Gender. Men tend to get a slightly bigger boost. Given what we explored in chapter 4, that men are generally less disclosive than women, it's likely that expressive writing is more novel and, therefore, more potent for them. If you rarely open up in daily life, being prompted to do so may pack a stronger psychological punch.

This difference is striking when I look back at my great-grandparents in the aftermath of Fred's death. Based on what I've uncovered, Nellie emerges as someone who talked and wrote about her grief—processing it with her son Bert, and likely others. Moses, by contrast, is largely absent from those accounts. I don't know whether he shared his feelings or not, but given the time, and the expectations placed on men, especially as head of the household, it's not hard to imagine he kept much of his grief inside. As we'll see, Nellie's approach is one we can all benefit from emulating.

Putting Feelings into Words

Anything that's human is mentionable, and anything that's mentionable can be more manageable.   —MR. ROGERS

It doesn't actually seem to matter how you put feelings into words— onto a page, a screen, or even spoken aloud. The point is to name what's swirling around in your head. Psychologists call this *affect labeling*. Think of it as emotional metabolism: turning raw feelings into words.

UCLA psychologist Matthew Lieberman and his colleagues demonstrated the power of affect labeling in a study with people who scored high on the "Spider Phobia Questionnaire" (yes, psychologists have a survey for everything). Participants were seated in front of a live tarantula (leg span: six inches) and given different instructions on how to cope. One group—the affect-labeling group—was told to try to put their fear into words ("I feel anxious the tarantula might jump on me"), while others were told to try different coping strategies, such as distraction.

A week after the initial tarantula experience, the researchers exposed participants to *another* tarantula. Then they measured participants' anxiety levels using that trusty measure of galvanic skin response, sweaty palms. Only the affect-labeling group showed reduced anxiety. In other words, it was only those who had practiced putting their fears into words who ended up coping better.

Intriguingly, this all happened in isolation. No audience, no therapist, no friend listening. Simply saying "I'm terrified" out loud was enough to make people feel less terrified. So if you ever feel ridiculous when journaling about your feelings, just remember that somewhere out there, a group of people stared down a massive spider, said "I'm terrified," and then when they saw another spider, walked away from it *less* terrified. Science for the win.

The apparent simplicity of affect labeling belies its neurological complexity. Imaging studies have shown that it changes the brain. It reduces activity in the amygdala, the mind's emotional alarm system, and boosts activity in the brain's regulation center that is linked to reasoning and control. Putting feelings into words literally calms the mind.

Affect labeling is a skill, and like any skill, it gets easier with practice. Most of us, for example, use too few words to describe our emotions. It's like painting with only two colors. Expanding your emotional vocabulary adds shades and detail, letting you identify what you're actually feeling. So instead of defaulting to broad strokes like "I'm angry," you might say "I feel aggravated that my spouse isn't listening to me—and *that's* what's making me angry."

I learned to do this from experience. The first time my therapist asked me to describe how I felt, I listed thoughts, not feelings. He finally handed me a "Wheel of Emotions," an emotional cheat sheet first conceptualized in the 1980s and since adapted many times. It helped me expand my emotional vocabulary and name what I was feeling.

Wheel of Emotions

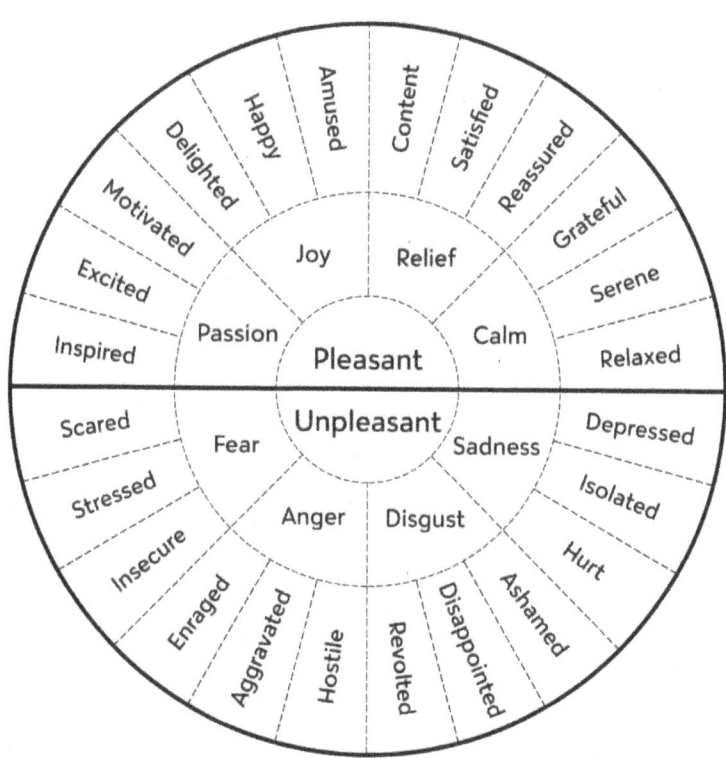

The wheel guides you to start with a basic emotion and refine it outward. For example, imagine you wake up feeling crummy. Step one is recognizing that the feeling is unpleasant. Dig deeper, and you might see it's sadness about a conflict with your boss. Go deeper still, and understand it's isolation—feeling you're on your own at work. These realizations can calm you and even spur action. Affect labeling isn't just about slapping a word on a feeling; it's about naming and understanding it with precision.

Armed with a better understanding of affect labeling, I went back to a letter Nellie had written to her son Bert, my grandfather, soon after she learned that Fred was missing but before she had received official confirmation of his death. I was curious to see how much emotion she had expressed. Quite a bit, though in a metered way, reflecting the norms of her time. Take this line: "It is getting us down though[,] the suspense of it, and the letters are torture." She was referring to the letters they were still receiving from Fred, written before he had gone missing. The pain is clear. But then she pivots to hope and the small actions taken to cope: "We are saying lots of prayers for him and also for the rest of you."

In another passage, she acknowledges the uncertainty head-on: "Poor darling Fred if we only knew . . . how it all happened." But even that is followed, almost seamlessly, by a return to daily life: "Dad received his usual box of chocolates [for Father's Day] and I helped him to devour them . . . we still have a few left." This weaving together of feeling and function, naming emotion while continuing to engage in ordinary life, is a productive way of coping.

CBT, Dodos, and the Talking Cure

Something else is worth noting about Nellie's writings: She wasn't just writing to herself; she was also writing to her son Bert. Writing to oneself can be helpful, as Pennebaker's studies showed. But it can be even more powerful to write to someone else—or, more aptly, to open up to someone else. As emotions scholar Amit Goldenberg told me, "Our minds are chaotic swirls of feelings. But when we try to describe them to others, it forces a kind of structure—a kind of logic—which, in and of itself, can be tremendously helpful." In other words, healing doesn't always happen in solitude. Sometimes we need someone else's mind to help make sense of our own. That's where talk therapy comes in.

Talk therapy has evolved considerably since the early ventures into what Josef Breuer and his buddy Sigmund Freud called the "talking cure," which they claimed to have used to cure hysteria (they didn't, because hysteria isn't a real thing). But the core question remains: What, if anything, does talk therapy do for us?

If you've looked into it, you've probably come across cognitive behavioral therapy (CBT). It's the most researched and widely practiced approach. The idea is simple: notice unhelpful thought patterns, challenge them, and replace them with healthier ones. Basically, retrain your brain to stop being its own worst enemy.

It's been tested again and again in rigorous scientific trials, and the verdict is clear: CBT works. But it's not a magic bullet. Its effectiveness ranges from modest to very strong, depending on the condition. It's especially good for anxiety, obsessive-compulsive disorder, and anger. For depression or post-traumatic stress disorder, it works best when combined with other treatments.

Other types of talk therapy help, too. In fact, most evidence-based approaches are about equally effective, especially compared to no treatment at all. This universality is known as the "Dodo Bird Verdict," a nod to a moment in *Alice's Adventures in Wonderland*. If you haven't read it, the gist for our purposes is that after Alice and a group of talking animals get soaked in a pool of her tears, the Dodo—an extinct flightless bird with a flair for absurd solutions—suggests a "caucus race" to dry off. There are no rules; everyone just runs around randomly until they feel dry. And then? Everyone wins. Just as the Dodo declared "all have won, and all must have prizes," psychotherapy research suggests something similar: Most evidence-based therapies are helpful, regardless of which route they take.

A Little Help from Your Friends

It's not just therapists who can help us open up. Sometimes the people who know us best make us feel safest to begin doing so. Friends and loved ones can validate our feelings—acknowledge them as real—and that alone is very comforting. I imagine this helped my great-grandparents cope in the days after the news of Fred's disappearance. They spent many hours with their old friend Debney, talking, remembering, sometimes just sitting quietly. That companionship, I suspect, was a big comfort.

In addition to validating our feelings and providing social support, friends can also help us to reframe or reinterpret what we're going through so it feels less overwhelming. This process is called *emotional reappraisal.*

Ever had a friend talk you down from an emotional spiral? They may well have been engaging in emotional reappraisal. Here's some evidence of its benefit: In one study, pairs of friends came into the lab. One person, the "experiencer," looked at upsetting images, like a car crash scene. Sometimes the experiencer just looked at the images. Sometimes they were told to try to reappraise the images themselves—for example, to tell themselves something like: "It looks bad, but help is already there." And sometimes they listened to their friend's reappraisal of the exact same images.

Afterward, the experiencers rated how badly they felt. Doing nothing felt the worst. Trying to reframe the images alone helped a bit. But hearing a friend provide effectively the same reappraisal was more helpful. In other words, *social reappraisal*—hearing someone else reinterpret the situation—was more powerful than going through it alone.

Scholars are still pinning down exactly why social reappraisal is helpful, but it's likely that putting emotions into words helps us step outside them, and having someone else do this makes the reframing easier to believe. It's one thing to tell yourself "This will pass." It's another to hear it from a friend or loved one.

Speaking of which, there are different ways of reappraising a situation, and researchers are beginning to discover which types are best. In addition to *temporal distancing* (e.g., "The misery won't last"), there's also *bright-siding*, which would be like when a friend tells you "At least something good came out of it." There's also *perspective-taking* ("Maybe the other person had their reasons").

Each of these can make you feel better than when your friend says nothing at all, but one of them is particularly helpful: temporal distancing, or the classic "this too shall pass" approach. This also happened to be the singular piece of parenting advice my cousin Sue gave me before I had kids. I found myself turning to it again and again (usually in the middle of the night, as during that parental rite of passage when your sweet baby projectile vomits directly onto your face).

But sometimes friends aren't around, or a cousin's sage advice isn't accessible to your sleep-deprived brain. That's led some, myself included, to wonder: Can AI be a partner in social reappraisal? It turns out that even generic, scripted reappraisals can help people regulate their emotions. This suggests that the benefit may come as much from *receiving* a reappraisal as from who delivers it. In theory, AI could be very helpful, providing on-demand support, always available and endlessly patient. But it's not without risks: Many tools have murky data policies and little clinical oversight, so privacy and safety are big open questions.

Lowering the Bar for Opening Up

Social psychologist Bernard Rimé has done extensive research on how emotional disclosure helps us cope. In one study, his research team showed participants three videos: a neutral nature clip, a cock-fighting scene, and an even more disturbing clip of people killing a monkey and eating its brains. Only the most disturbing video made people feel compelled to talk about it. They concluded that the urge to share socially kicks in only once an intensity threshold is crossed.

That made me wonder whether we might benefit from lowering this threshold. Do we make a mistake by staying silent about every-day struggles—work stress, relationship worries, lingering anxieties—the ones that aren't extreme but still wear us down?

In a study I conducted with Mario Small, Kristina Brant, and Ximena Garcia-Rada, we asked a nationally representative sample of 1,099 people to list their personal struggles. Their answers ranged from financial worries ("Been unemployed for seven months") to fears about health ("Tested positive for the BRCA gene—breast cancer likely imminent") to relationship turmoil ("Husband had emotional affair, divorce options"). We then had them list the seven people closest to them and asked whether they had shared their concern with anyone on that list. We found that people often kept their problems to themselves; it was common for them to not have confided in any of their seven closest people. But those who had confided were significantly happier overall.

Intrigued by this result, we wondered whether "prescribing" disclosure—tasking people with confiding in a close other—might be helpful. In a second study, we randomly assigned half of the participants to talk to someone close to them about an issue they were facing. They could choose anyone they wanted. Two days later, these people felt better about the issue and were happier than those who we hadn't prompted to share.

In a final study, we pushed things even further. It was the same setup, except that instead of letting the participants choose whom to confide in, we had them first list their seven closest contacts, and then we (randomly) picked which of these contacts they were to go confide in. We found that this, too, was helpful: These people felt better two days later relative to people who hadn't been prompted to share. This suggests that when something is weighing on us, we don't need the perfect confidant—we just need to talk to someone who cares.

Crossing the Openness Barrier: The Secret History of MDMA

By now we've seen that self-disclosure is central to emotional processing and well-being. Whether through expressive writing, affect labeling, or talking to someone, putting feelings into words helps us metabolize them, making it easier to cope. But what if the things you hold on to feel impossible to say?

For decades, scientists, therapists, and underground psychonauts (!) have wondered if certain psychoactive substances might lower barriers to disclosure. Could a carefully guided experience with a hallucinogen or empathogen help people access and articulate what they otherwise suppress?

Recall our gregarious psychologist Sidney Jourard, the self-disclosure expert we met back in chapter 4. He also dabbled in drugs and on occasion experimented on himself. His substance of choice? LSD. Once, he used a projective test called the "Incomplete Sentences Blank," which has you finish stems like "My greatest fear is . . ." or "I am best when . . ." He filled it out before the acid trip and again four hours in. Here are some of his responses (you're welcome):

My greatest fear . . .
Pre-LSD: ". . . is that I'll lose a court case over my dog and have to pay $100."
During LSD: ". . . is fear of fear—yet when I face fears, they no longer frighten."

I can't . . .
Pre: ". . . wait to finish my book."
During: ". . . be all the things I'd like to be. More's the pity."

The future . . .
Pre: ". . . will probably be all right, though I feel doubts about it."
During: ". . . is uncertain, but isn't that the fundamental property of futures?"

I need . . .
Pre: ". . . a congenial job with a good salary, stimulating colleagues."
During: ". . . more salary so I can forget it, & do productive work, which I love to do."

I am best when . . .
Pre: ". . . I am relaxed."
During: ". . . I feel loved."

Is this scientific? Nope. But it's fascinating, right? Jourard's "before" answers were mundane; his "during" answers were abstract, self-aware, and emotionally honest. Jourard's trip points to how altered states can make us more open and reflective. Which raises a bigger question: Could certain drugs help unlock disclosure?

One candidate is MDMA, often sold in impure form under the street names Ecstasy or Molly. Its story begins not in nightclubs or therapy rooms, but in a German lab in 1912.

That year, Anton Köllisch, a chemist at Merck, first synthesized 3,4-methylenedioxymethamphetamine, now known as MDMA. He had been working on a blood-clotting drug, and MDMA turned up as an accidental by-product. Because MDMA wasn't the intended medicine and didn't seem useful, Merck shelved it. It was never tested with people.

Decades later, during the Cold War, the CIA swept MDMA into its infamous MK-ULTRA experiments, a sprawling program that tested dozens of substances to see if any might serve as a "truth serum"—something to make suspects pliant and reveal secrets. But MDMA didn't fit the bill. Rather than making people compliant, it made them relaxed. So MDMA was shelved again.

It wasn't until the 1970s that MDMA sprang up again, synthesized this time by renegade chemist Alexander Shulgin. An inventive and unconventional figure, Shulgin was known for self-experimentation and for sharing psychoactive compounds with therapists and researchers. He quickly recognized that MDMA wasn't just another stimulant. It seemed to strip away fear and defensiveness, making emotional access easier. He called it "window," because it offered a clear view into the self, free from shame and self-judgment. And when partaking himself, he was known to have referred to it as his "low-calorie martini."

Later animal studies support these ideas. Rats given MDMA exhibit "adjacent lying" and "prolonged snout-to-snout contact," the rat equivalent of cuddling. In primates, MDMA increases grooming behaviors, an indicator of trust and connection. But maybe the most intriguing study was the one with octopuses. After receiving MDMA, they actively approached each other, even wrapping tentacles. This was an astonishing shift for a species that typically avoids social engagement outside of mating.

In 2021, a landmark double-blind, placebo-controlled study tested MDMA-assisted therapy for PTSD. By the end of treatment, only 33 percent of those who had received the drug still met the criteria for PTSD, while 68 percent of the placebo group still did. Though these results seem strong, the FDA aptly declined to approve MDMA therapy for PTSD out of concerns about scientific rigor and long-term safety. That said, the promise of psychedelic-assisted therapy is hard to ignore, given that it appears to facilitate self-disclosure.

Indeed, many MDMA trial participants describe the experience as a breakthrough in emotional openness. One called it an "evaporation of the usual barriers to intimate communication." Another said it helped them "experience the ease of expressing myself." Neuroscience helps explain why: Like affect labeling, MDMA quiets the brain's fear center and boosts regulation from higher-order brain regions. It also floods the brain with serotonin and oxytocin, chemicals tied to bonding, safety, and trust. Some researchers even speculate that MDMA and affect labeling could work together in clinical settings, one lowering the fear threshold, the other helping to structure what is revealed. I'm excited by the possibility, though we still need more evidence before drawing firm conclusions.

The Joy of Self-Disclosure

So far, we've been focused on the healing power of opening up. If you're the kind of person who finds it hard to do that, it may sound about as appealing as doing burpees. But suppose I tell you that it's not like that at all. That once you start to do it, it feels great. I don't think that's entirely coincidental. Nature has a way of making things that are good for us pleasurable.

The pleasure—and relief—of opening up explains the popularity of games like Twenty Questions, Never Have I Ever, and Two Truths and a Lie. All revolve around the reality that there is so much that even those close to us don't know about us. The pleasure of that disclosure explains why we play, despite knowing we might end up revealing something ridiculous we've done. Like that time I told some colleagues that I had taken a giant bite of my succulent-orange-scented bar of soap in the shower. It smelled so convincingly real that I could have sworn it would taste just as it smelled (those of you who have ever tried the Body Shop's Satsuma scent just might empathize with me). And yes, I was well above the age of six when I did this.

Neuroscience backs up the soap story. Self-disclosure can trigger the same biological hit of pleasure that children get from being found in hide and seek. In one study, Diana Tamir and Jason Mitchell scanned participants' brains in a functional magnetic resonance imaging (fMRI) machine. Half the participants answered questions about themselves (such as their favorite ice-cream flavor) while a friend listened. The others listened to their friend answer instead. The "pleasure centers" of the brain lit up more during self-disclosure than mere listening.

We like sharing so much that a surprising percentage of our everyday conversations is about ourselves—including trivial "fun facts," like our favorite ice-cream flavor. Tamir and Mitchell found that people were even willing to give up money just to keep talking about themselves. People gave up an average of about 17 percent of their study earnings to do so, rather than switch to another task.

When this finding came out in 2012, it rocked the world of privacy researchers to which I belonged. You see, I'm somewhat of a "recovering privacy expert." I used to study disclosure decisions from a very different lens than I do now: from the perspective of behavioral economics, which is fascinated (read: borderline obsessed) with the seemingly misguided, irrational decisions we make. And, in studying decisions surrounding privacy, we'd hit the mother lode. It was astounding to us how comfortable people seemed to be with pouring their hearts out online, despite the many risks.

Around that time, it became clear that this wasn't just a laboratory curiosity. The late 2000s made "oversharing" a household word (coined by Emily Gould, who later regretted her own tell-alls on *Gawker*). Examples abounded. There were the flight attendants who were fired after venting online. There were the early blogging platforms overflowing with jaw-dropping confessions—everything from having sex with biological family members to cheating on a dying partner to removing a ten-day-old tampon. Offline, too, the memoir boom (up 400 percent between 2004 to 2008) showed how hungry people were to reveal private struggles. In hindsight, those public confessions were less a cultural anomaly than a case study in human nature. They reflected the same reward-seeking impulse that Tamir and Mitchell had uncovered in the lab.

In fact, Tamir and Mitchell's research suggested that people may get so much biologically measurable pleasure from sharing that it overrides their caution about what the consequences might be. I came to this same realization, albeit in a more roundabout way.

When I was in graduate school, I worked with two renowned economists, Alessandro Acquisti and George Loewenstein, on a study we cheekily titled *"How BAD Are U???"* (HBRU, for short). We built two versions of the same nosy questionnaire. One was stripped down and looked relatively professional:

Survey of Student Behaviors

4. Have you ever smoked marijuana (i.e. pot, weed)?

○ Yes
○ No

5. Have you ever "cheated" while in a relationship?

○ Yes
○ No

6. Have you ever driven when you were pretty sure you were over the legal alcohol level?

○ Yes
○ No

The other one looked like this:

How BAD are U???

4. Have you ever smoked marijuana (i.e. pot, weed)?

○ Yes
○ No

5. Have you ever "cheated" while in a relationship?

○ Yes
○ No

6. Have you ever driven when you were pretty sure you were over the legal alcohol level?

○ Yes
○ No

For months I lugged my hefty laptop around Pittsburgh to show people the two sites and ask them which one seemed safer (the glamorous life of a behavioral scientist). As we expected, almost everyone pointed to the professional-looking one as being far safer. The whole point was that HBRU looked like a dangerous site—exactly the kind of place you should not share.

Yet when we asked people to answer questions on one or the other site, people spilled far more on the gonzo site, though it looked incredibly sketchy, down to its comic pixelated red devil icon and eccentric spelling. For example, 17 percent admitted to drunk driving on the professional site, compared to 30 percent on HBRU. Twelve percent confessed to peeping on someone undressing on the pro version, versus 20 percent on the devil site. Twenty-two percent admitted to cheating on a partner on the clean site; 31 percent did so on the devil one. What about trying cocaine? Two percent on the pro survey, 7 percent on the other.

At first this baffled me. Yet more proof of how bizarrely irrational the human species is, I thought. But over time, I came to a different interpretation. That's because whenever I presented the results, audiences always laughed when I showed the devil site. Eventually, I realized: Maybe people disclosed more not despite the absurdity, but because of it. The site felt like a joke, which made the confessions oddly fun—even though many of the questions asked about behavior that was clearly wrong, even illegal. Looking back, that tension—the humor of the site versus the seriousness of what people shared—has stuck with me. It's a reminder that playfulness can lower inhibitions, for better and for worse.*

Presenting these findings made me rethink the "problem of shar-

---

*This study was conducted nearly twenty years ago. If I were designing it today, I'd design it differently; I'd tweak it to avoid inadvertently implying that serious or harmful acts are trivial.

ing." Because while in my professional life I was wagging my finger at people for oversharing, in my personal life, I was happily sharing away—filling out *BuzzFeed* quizzes (my kryptonite) and forking over Facebook data to find out what my baby would look like if I married a celebrity. Did I regularly accept cookies (the tracking kind) online and maybe to this day keep important passwords on a Post-it? Am I not known by friends and colleagues as the most likely to say too much? (Full disclosure: I've spoiled not one, not two, but three surprise birthday parties. So far . . .) I have read, amassed, and analyzed more than enough data to know that my behaviors are risky, but that still hasn't changed them much. Why, then, am I so bad at privacy? And why am I not ashamed of this? Why am I even telling you this?

Many research projects and a good deal of soul-searching later, I can say that, yes, having fun was a big part of the HBRU story. One of the main motivators of sharing personal information online—and beyond—is the sheer pleasure we get from doing so, and I think the gonzo look of the site amplified the fun.

But as I thought about our findings, I realized the deeper takeaway was that many of us would like to reveal more about ourselves, including embarrassing things we've been told to keep private. For the participants in our study, it seemed that revealing so much about behaviors they felt they should keep secret was a source of relief from the constraints most of us feel against self-revelation. Maybe sharing was a way of letting go of shame.

As we discussed earlier, we're good at keeping secrets, but these secrets weigh on us. When we gave our participants the chance to get some of their secrets off their chest, many of them went for it. Sharing our secrets—whether with a trusted friend, family member, therapist, or even on a ridiculous-looking survey, can be incredibly freeing.

On May 8, 2003, sixty years to the day after my great-uncle Fred Duquette went missing, my parents stood at the south-eastern tip of Malta, looking out over the sea. They had come to honor him.

It had been pouring all day. But as they neared the place where the search for him had ended, the rain began to lift. The sky cleared, just enough. It felt like something had shifted.

Maybe that's what revealing does. It doesn't fix the past. But it helps us to carry it, with more clarity and maybe a little less weight.

Sometimes healing takes time. Sometimes it takes generations—preserved in letters, shared in conversations, and finally let go, little by little, in the telling.

And then at 6:35 p.m.—the exact minute Fred's plane had vanished from the sky—a rainbow appeared.

7

Building Friendship

When I teach people about the power of revealing and the rewards they stand to reap by being just a little bit more open, I start the class with an exercise. I tell the class that in a moment, everyone is going to turn to their neighbor, whom they don't know, and chat with them for five minutes on the same topic—a topic that I will supply. I tell students on one side of the room to close their eyes; the other half watches as I write their topic on the board: "What do you like about your job?" Then I quickly erase it, and I tell this group to close their eyes while those on the other side of the room open theirs. I write this group's question on the board: "When was the last time you cried?" Sometimes an audible groan rises up in response to this question. This isn't surprising. Talking about the last time you cried consistently ranks dead last on the list of topics people want to share with a stranger or an acquaintance, or even a loved one, for that matter.

Then I give everyone five minutes to answer their assigned question with their neighbor.

I walk around the room observing and listening. The result is always striking, and it's always the same: One side of the room is completely energized, with people leaning toward each other and making eye contact, while those on the other side look like they're suffering through an 8:30 a.m. chemistry class. Snoozer.

After the exercise, I ask people to raise their hand if they'd like to repeat this exercise (making a jokey quip for their partners to not take it personally). Hands shoot up for half of the class. And it's the criers who put their hands up, despite the fact that many of them groaned or even rolled their eyes when I asked them to talk about the last time they cried.

But when I assign them to open up, by the powers vested in me as a professor, people love the experience. They walk away from class with new connections and the spark of new friendships. This simple exercise is a demonstration of what hundreds of studies have shown. Mutual self-disclosure is a powerful impetus for friendship.

To understand why, we have to consider the two main reasons that revealing leads to liking and the possibility of friendship. First, revealing signals that you are willing to take the risk of trusting the other person with personal, and perhaps even embarrassing or shameful, information about yourself. It's quite possible, after all, that the other person will respond to your revelation with shock, disgust, contempt, or any number of negative emotions. Choosing to open up to them anyway signals that you trust them, which begets their trust in you—and their liking of you.

The second reason that revelation leads to liking and friendship has to do with the substance of what we share—specifically, whether it is something that builds common ground. When someone reveals something we really relate to, we feel closer to them. We see a kindred spirit. Finding that you have common beliefs, feelings, or even quirks goes a long way toward building liking in the early stages of friendship.

There's one more twist to my in-class exercise. After I reveal that each side of the room had a different question, invariably the side of the room that got the boring question feels jilted. So, in a closing flourish, I announce that I'll give one more topic for discussion, framing it as providing an equal opportunity for people to forge connections. I tell them that once I put it up, they're to turn to the person on their other side and answer it. Then, very matter-of-factly, I go to the chalkboard and write: "Please describe any sexually transmitted diseases (STDs) you have had, past or present." Then I wait. I give it a good five seconds. Students shake their heads. "No, no, no," some say under their breath. "But that's so inappro—" exclaims one student. It's a little indulgent, but yes, I revel in the reactions for a few moments. Then, just before a pending mass defection, I say, "Just kidding!"

Nervous, then legit, laughter ensues. Obviously, *this* is TMI. And that's the point. In addition to seeing how opening up about sensitive things, like the last time you cried, is bonding, students *also* see how it's possible to overdo it. Achieving the balance between TMI and TLI is vital when forging friendships. But finding that Goldilocks spot can be especially daunting in a new friendship, where we lack the familiar contours of an established relationship to guide us. Fortunately, a deeply ingrained instinct—reciprocity—can guide us.

Reciprocity

Humor me for a moment and imagine we're traveling back in time together to the early days of human life. We belong to a tribe of hunter-gatherers, and for us, survival is a daily puzzle. Some days we return from the hunt empty-handed; on others, we're lucky— someone in the group manages to bring down a large animal, and suddenly we have far more meat than we can eat before it spoils.

So what to do with all that excess meat? We could hoard it, sure, but then it would rot—and in a few days, we'd be hungry again. Or we could share it with others in nearby groups we sometimes have contact with, knowing that one day *they* might get lucky, and we might come home empty-handed. By sharing in times of plenty, we increase the odds that others will share with us when we're the ones in need. Even without formal agreements, this kind of give-and-take, go-with-the-flow generosity creates a safety net for everyone. It also just feels . . . right.

Anthropologists and evolutionary psychologists have proposed that precisely this kind of feast-or-famine dynamic—especially around high-risk, high-reward pursuits like hunting—may have given rise to the norm of reciprocity, one of the most powerful forces in human social life. In an influential 1960 article, sociologist Alvin Gouldner defined the norm of reciprocity as making "two interrelated, minimal demands: (1) people should help those who have helped them, and (2) people should not injure those who have helped them."

This norm may even have shaped aspects of human evolution. Zoologist Matt Ridley suggests that our instinct for self-preservation motivated us to become more cooperative: By developing a norm of reciprocity with other groups, we helped ensure our own survival. Because reciprocity promotes group cohesion, risk-sharing, and trust, it may have been favored by natural selection over time, so much so that it now feels like second nature. Many primates show signs of reciprocal behavior—monkeys groom those who groom them, for example. But in humans, reciprocity became more formalized and far-reaching, extending beyond kin, across time, and into abstract moral norms. Paleoanthropologist Richard Leakey and science writer Roger Lewin even argued that reciprocity is one of the defining features of human social life, not because it's uniquely human, but because of how deeply we've embedded it into our cultural and moral fabric.

Of course, humans don't just share meat. In most foraging societies, the majority of calories come from gathered plants and small game—resources that are typically shared, too. But it's those occasional windfalls, like a massive animal kill, that may have pushed our early ancestors to formalize sharing rules that still shape us today.

We experience this built-in impulse toward reciprocity routinely in everyday life. If a friend buys lunch, we offer to "get it next time." If you shovel your neighbor's sidewalk, they might bake you cookies. Failing to reciprocate makes most of us feel guilty or ashamed. And when our own kindness goes unacknowledged, we notice. You might not expect cookies, but it stings a bit if your neighbor doesn't even mention your help.

I first found myself caught in a cycle of mutual obligation at age six, when my brothers and I got into a "battle of the nice guys" with the elderly couple who lived in the flat below us in Freiburg, Germany, where we were living at the time. It all began one day when we arrived home to find a treat on our doorstep: a bag of Haribo gummy candies. Our parents encouraged (ahem, told) us to write thank-you cards; we obliged.

In hindsight, these cards—hand-drawn masterpieces on my dad's dot matrix printer paper—came across as a return gift. After all, gummies are to kids what hand-drawn thank-you cards are to the elderly. So, a few days later, another bag of candy appeared on our doorstep. More thank-you cards were manufactured. And on and on, until it got a bit awkward for us. We felt our thank-you cards may have been making the couple feel obligated to keep giving us gummies. So, after a fourth bag of candy appeared on our doorstep, we restrained ourselves from writing a note, though it felt weird. And to everyone's relief, that was the end of our gummy bear gravy train. I was reminded of this early-career case study in reciprocity when my brothers and I finally—we're all in our forties!—pledged to stop giving one another birthday gifts. Not having to stress our-selves out about these presents anymore was its own kind of gift to one another. (I reserve the right to send cards, though.)

Yet giving and receiving isn't only about social cohesion (and candy). Gouldner argued that in addition to helping society run smoothly, reciprocity jump-starts our social interactions. And for our purposes, that includes friendship. The person who takes the first risky step of offering the gift of disclosure builds trust and—hopefully—prompts a disclosure of similar magnitude from the other side, especially if the other person is motivated to forge a new friendship.

As famed psychologist Arthur Aron put it, "One key pattern as-sociated with the development of a close relationship among peers is sustained, escalating, reciprocal, personal self-disclosure." Aron and his colleagues wanted a reliable way to spark connection in the lab, so they designed a protocol that would manufacture friendship on de-mand. (Sorry—that sounded a bit clinical.) Strangers are paired up and asked to take turns answering a series of thirty-six increasingly personal questions. The early questions are pretty light, including "What would constitute a 'perfect' day for you?" and "Would you like to be famous?" Gradually, they deepen: "What roles do love and af-fection play in your life?" "Share with your partner an embarrassing moment in your life," and "When did you last cry in front of another person?" Compared to small talk, people assigned to do this struc-tured exchange like each other more, enjoy the task more, and feel more connected to each other (much like my students, who have to discuss only one of these thirty-six questions). Though the task has come to be known as "the path to love," and it's the subject of lots of magazine and newspaper articles about how to foster romance, it was developed as a path to any kind of close relationship.

One clever twist on Aron's study reveals something else that matters: the sequence of mutual disclosures. In a variation of the experiment, some pairs were instructed to take turns answering questions while others were told to have one partner go through all the questions first, with the other responding only afterward. The results? The turn-takers felt more connected and enjoyed the experience more. The sequencing mattered, not just the content.

This deceptively simple tweak taps into something intuitive: Turn-taking makes vulnerability feel safer. We want to escalate, but *together*. Especially at the beginning of a friendship, we're uncertain about how someone will respond to our disclosures. When someone matches our openness right away, it signals safety—and invites us to keep going. In this way, early friendship is less like a lecture and more like a conversation. We reveal, we receive, we respond.

Think back to your first real conversation with someone who ended up becoming a close friend. You know that feeling when you really click with someone? Maybe, whether over dinner, on a walk, or at a study session, you started by sharing information about neutral, unemotional topics, like what you were planning to do that weekend or how much work you had to get done before then. Then, gradually, the silences grew shorter as you shared more about your lives and discovered you had things in common: a love of long, aimless road trips with overambitious playlists; or that you both grew up in houses where the thermostat was treated like a sacred artifact—touch it and suffer the consequences. Maybe you confessed that after all these years, you still harbor an irrational (or rational?) fear of ventriloquist dummies. Or maybe you both admitted to organizing your books not by author or topic but by color. (All of these are true for me!)

Such disclosures then went deeper when one of you revealed

something more personal. Perhaps your new friend shared that their mom was battling breast cancer, and you said you had a cousin going through the same thing. Or maybe you alluded to an issue you were having with your romantic partner—just testing the waters to see how they'd respond—and the other person, to your relief, reciprocated with understanding and a sensitive revelation from their own relationship. The time flew by, and before long you were laughing, your eye contact sustained, your optimism growing that you had found someone you could be truly yourself with.

What matters in these moments isn't so much that you both talk about the same thing. It's that you both reveal something equally personal and real. Matching the emotional depth tends to matter more than matching the topic—especially when the response feels attuned and affirming. You could both be talking about family and completely miss each other on depth—one person describing a favorite recipe from their childhood, the other revealing how hard it's been to care for a parent with dementia. Conversely, you might be talking about entirely different topics—say, one person opening up about feeling unmoored after a recent breakup, the other sharing how unexpectedly emotional they felt watching their child walk into kindergarten. Different stories, same depth. It's not the topic that builds connection, per se. It's vulnerability.

Mutual disclosure that is balanced in vulnerability is so powerful a catalyst that it even shows up in our interactions with computers. In a 2000 study run by my treasured colleague Youngme Moon, participants who first saw a computer that was programmed to reveal information about itself were more likely to share something personal in return. In one case, the computer said it "rarely gets used to its full potential" because 90 percent of users don't use applications that require its top speeds. Then it asked, "What has been your biggest disappointment in life?" Participants who saw the computer's disclosure before this question were more likely to share regrets than those who didn't. People also rated the computer as more likable when it revealed something they perceived as vulnerable. This was true even though Moon took care to avoid misleading participants into believing the computer had feelings. The messages were text-only, used prewritten scripts, and the computer referred to itself as "it" rather than "I." It seems we humans are so drawn to mutual openness that we respond to it—even when it comes from a machine.

You can think of the role of self-disclosure in sparking friendships like building a fire: You first gather up the kindling—proximity, social curiosity, openness, extraversion, questions. The spark that starts the fire can be some point of commonality you've noticed. But even with kindling and a spark, you still need one more thing for a fire to catch: oxygen. And in the world of friendship, that oxygen is attention and reciprocity.

Four Ways the Fire Falters: Reciprocity Fails

When I was a baby academic, attending a recruiting conference in the hopes of landing my first professorship, I was sapped of energy. I was more than a little tired of putting on a happy face in interview after interview. At one point, I found myself alone in the elevator with another young academic who was also on the job market—which we both immediately recognized, thanks to our brand-new, slightly ill-fitting black suits (the rookie job market uniform of choice).

I looked the guy in the eyes. He looked drained, too. I took a big breath, exhaled with my whole body, and said, "I'm so exhausted."

Instead of reciprocating, he launched into the canned interview song and dance: "I find it exciting. It's such an amazing opportunity to meet so many esteemed professors and get such valuable input on my research agenda."

Seriously? Couldn't we be real for just this one moment?

"Uh-huh," I mustered as I stared ahead and rolled my eyes. The doors opened, and we went off to our next interviews.

We were both fortunate to get great jobs. We see each other at conferences semiregularly, but we've never become friends.

As my brief elevator encounter suggests, the sting of a reciprocity fail lingers.

Reciprocity falters for many reasons—too little, too much, mistimed, or one-sided—and when it does, the fire between people starts to sputter.

Misfire #1: The Fizzle

The most jarring kind of reciprocity fail is of the TLI variety I experienced on the elevator, when someone simply doesn't reciprocate after you've made yourself vulnerable via self-disclosure.

It's rare for openness to be met with silence, but that can happen. More commonly, the way this failure manifests is with your would-be friend failing to open up to a commensurate degree. They might say something superficial in response, change the topic to something less personal, or, my personal pet peeve, relentlessly ask you questions to keep the focus off themselves (you know who you are!). Whether intentional or not, this sends a strong signal: The fire's not catching. Translation: A new friendship isn't happening, at least not in this moment. It's like laying down the first bit of kindling, hoping for a spark, only to find damp wood in return. Without a mutual offering, there's nothing to feed the flame.

The sting of a reciprocity failure tends to stick with us. Yes, sometimes it really is something we said or did. But often, such seeming "rejection" has nothing to do with us, or with them, for that matter. As psychologist Lee Ross pointed out in his research on the "fundamental attribution error," we're prone to blaming others' personalities for behavior that's actually situational. Maybe your would-be friend was exhausted, distracted, or just not in a place to connect. A nonresponse isn't necessarily a rejection. I prefer to think of it as a sign to take my kindling elsewhere.

Take the "I find it exciting" elevator guy. Perhaps he was so dialed into the interview process that he just didn't have it in him to let down his guard at that moment. Or maybe I incorrectly thought I saw my own fatigue mirrored back at me—maybe he truly was full of energy and excitement, and trying to buck me up. If so, maybe he thought *I* was the one who dropped the friendship ball! As hard as it is for me to accept it, there are many possible reasons unrelated to his character for why he didn't respond as I'd hoped.

It isn't necessarily a bad thing when people don't reciprocate, or they reciprocate in a way that is off-putting (even though it can be uncomfortable). On the contrary, it gives you valuable information. It tells you that, for whatever reason, a friendship with this person, at least right now, isn't going to catch. Just like in dating, a lack of reciprocation helps you avoid wasting your time and frees you up to look for a better match.

Misfire #2: Choking the Fire

At the other extreme, we can fail in the mutual disclosure process by overreciprocating, too. When someone reveals something to you, and you see friendship potential, the principle of reciprocity obliges you to respond with something that is at least as sensitive as what your would-be friend has shared. You want to see them, and maybe even raise the ante a little bit. But if you go all in, you risk coming across as maladjusted, a hot mess, or just bizarre. Hardly friend material.

Take Lily, a partner at a law firm who ran into an acquaintance, Nia, at an art gallery opening. Nia was a friend of a friend; she and Lily had met a few times at parties. This was a nice time to start to get to know each other better, Lily thought, as they chitchatted over their champagne. Lily mentioned to Nia that she was excited about an upcoming trip to France, but anxious over leaving her young child with her mother. So far, so good: a small, appropriate escalation of sharing.

Now it was time for Nia to reciprocate. She responded by telling Lily about an experience she and her husband recently had as they were about to travel to Europe. They were flying standby, and were fortunate to get seats—but one seat was first class, and the other was in the back of the plane. "I immediately saw the negotiation angle," Nia declared proudly. "I said, 'If you give me the first-class ticket, I'll give you a blow job later.' And so a deal was consummated."

Ooookay.

Lily was taken aback. Now, I should say that Lily is not remotely prudish; coming from a good friend, the story would have amused her. And the desire to fly first class internationally *is* understandable. But the blunt sexual reference, coming from someone she barely knew, was TMI, too soon. And it certainly didn't encourage her to up the ante with any similarly revealing confidences! Lily nodded politely and soon ended the conversation—and the prospect of friendship.

It's a classic case of smothering the spark: piling too many logs onto a fire that hasn't yet caught, that hasn't had the chance—or the oxygen—it needs to breathe. The impulse might be warm, even generous, but the effect is smothering. And it doesn't always show up as a wild, unfiltered anecdote. Sometimes it's just the mismatch of intensity—too deep, too fast, without a sturdy base beneath it. You try to feed the fire, but you overwhelm it instead. Other times the problem isn't the heat but the rhythm of the fire-building itself. Which brings us to the next misfire: the moments we forget to pass the torch.

Misfire #3: Failing to Pass the Torch

A while back, I found myself chatting with a fellow mom at gymnastics. It started with the usual small talk: "You left the baby at home today! Nice break for you, I bet." She responded with a full-on trauma dump about her baby's severe reflux and all the measures she'd taken to get him to eat and gain weight. Though I empathized because I'd been in that position myself a year earlier, the conversation was completely one-sided. She hadn't asked a single question, let alone paused for me to say something. Realizing that she was likely in survival mode, I went into listening mode. Something vital had been skipped: the back-and-forth rhythm that helps a new friendship spark to life. The fire never moved beyond the first flame. She hadn't passed the conversational torch (completely understandably).

Remember that study I mentioned earlier about how much we all love talking about ourselves? That impulse can derail turn-taking. We get so caught up in our own story that we forget it's not all about us. A fragile ember flickers out before it has the chance to catch. Fortunately, there's an easy fix—ridiculously easy: *Ask a question.* Just one simple question can be enough to fan that sputtering fire back to life. Almost any question will do. It might feel awkward or overly deliberate at first, but your conversation partner probably won't think so. People love being asked questions.

Early in the pandemic, my lab instituted something we called "Zoom popcorn" during our virtual meetings. The game was simple: Answer a fun icebreaker—like "What's the most life-changing invention?" (my answer, fresh off giving birth, was epidurals)—then call on someone else. Most people managed this just fine, but I kept forgetting to pass the torch. I'd get so wrapped up in my own answer that I'd finish talking . . . and just stop. While I can conveniently blame the brain fog of new parenthood, the truth is, most of us, myself included, forget to pass the conversational torch more often than we realize.

Follow-up questions—those that build directly on something your budding friend just said—are an especially powerful means of fostering closeness. In research led by Mike Yeomans and Alison Wood Brooks (more about her soon), people who asked lots of follow-up questions were better liked. Why? Because follow-up questions signal that we're listening, we care, and we want to hear more.

And if you're worried that asking questions might come across as intrusive or nosy? According to Einav Hart and colleagues, we systematically overestimate the discomfort our questions will cause. Even sensitive questions—about money, politics, pregnancy—are often (though not always!) received better than we expect.

The key is simple: Don't keep the flame to yourself. Pass the torch.

Misfire #4: Mistaking Smoke for Fire

Sometimes we think we're nailing reciprocity when actually we're completely missing the mark. It happens when we mistake knowing someone for being known *by* them. In other words, we assume that social ties are mutual—that if I feel close to you, then you must also feel close to me. This is a product of a one-sided disclosure dynamic:

when we know a lot of personal things about someone, but they don't know anything about us, or even that we exist.

Let me make this more real with the embarrassing story I'm about to share, at my own expense, yet again.

My husband Colin and I were attending a play at Lincoln Center. In the lobby during the intermission, I spotted none other than Jerry Seinfeld, milling about with his wife, Jessica (yes, I'm a stalker girl who knows his wife's name . . . you can guess where this story is going). Adrenaline surged through my veins. I felt flushed. As it happens, I have a possibly unhealthy love of the show *Seinfeld*.

So, I sauntered up to the Seinfelds as if we were old friends. "Jerry!" I cried. "I love your—" But before I could finish my unoriginal compliment, Seinfeld's eyes glazed over and a giant, fake smile was suddenly plastered on his face. "Great to see you! Great to see you! Great to see you!" he said loudly and condescendingly, muffling the possibility of me saying anything more. He and his wife hustled away, fast. Like, as fast as you could and still count it as walking.

As you might guess, I was mortified. What an idiot I was! Just because I felt like I knew Seinfeld didn't mean *he knew me* or wanted anything to do with me. Even worse, this type of thing must happen to him constantly—so often he'd come up with an exit strategy: saying "Great to see you!" over and over while run-walking away.

Colin, who watched the whole scene unfold, still snickers at the memory. "It was even more entertaining than the play," he recalls.

In retrospect, it seems fitting that I had a run-in with my favorite show's star that was cringey and awkward enough to be worthy of its own *Seinfeld* episode. The fact that I'm not the only person who has made a fool of herself by assuming a false camaraderie with a celebrity provides only cold comfort. A friend of mine, for example, told me about the time he saw Jennifer Lopez sit for an autograph session at a venue in Harvard Square. As each person approached, her bodyguard told them: "You do not know Ms. Lopez. She does not know you. Do not touch Ms. Lopez. Do not ask her questions. You do not know Ms. Lopez." Where was that bodyguard when I needed him at Lincoln Center?

It seems many of us are susceptible to what I call the *illusion of reciprocity*—the tendency to assume that social ties are mutual: If I view you as a friend, then you must view me as one, too.

This illusion thrives in *parasocial relationships*. Coined in 1956 by Donald Horton and R. Richard Wohl, the term describes those one-sided bonds we form with celebrities. Today, that includes influencers, podcasters, and in my case, Jerry Seinfeld. These are people who are completely unaware of our existence. A classic example is from the 1950s children's television show *Ding Dong School*, hosted by Frances Horwich, known as "Miss Frances." Her gentle demeanor and direct address to the camera made children feel as though she was speaking directly to them. This connection was so strong that at least one child attempted to "free" her from the television by removing the back of the set. This prompted Horwich to do in-person shows to prove to children that she was not trapped in their TVs. As this early instance highlights, media can foster deep, albeit one-sided, bonds.

Today, the illusion of reciprocity thrives on social media. In and of itself, social media isn't bad for friendships: The one-on-one conversations that sprout up can birth and grow relationships, and renew dormant ones. However, two particular aspects of social media give rise to the illusion of reciprocity.

The first is that our online ties need not be mutual—that is, the people I follow don't necessarily follow me. And yet we implicitly assume that our "friends" follow us back, as a collaborator and I found in a set of studies. We asked participants to list a few people they follow regularly on social media. Then we asked them to indicate (1) how much they liked the people they followed and (2) how much the people they followed liked them. Not surprisingly, participants really liked the people they followed, but they *also* said that the people they followed liked them as well! As a reality check, we then asked participants whether those people followed them back—arguably a low bar indicator for mutual regard, or even basic acknowledgment of their existence. Of course, they rarely did. Yes, the illusion of reciprocity can veer into the outright delusional.

The second aspect of social media that fosters this illusion is the simple fact that it is often an orgy of TMI, since the algorithms reward engagement, and let's face it, our curiosity for stars' intimate self-disclosures is nigh insatiable. Think Will and Jada Pinkett Smith sharing gory details about their marital ups and downs, or the Kardashians just living life out loud day after day. As a result, we can wind up knowing a lot about someone, which leads us to like them and feel close to them despite them not knowing anything about us.

In everyday friendships, this misstep shows up in subtler ways. We may believe we've opened up because we've talked a lot, but if that talking isn't mutual, we may be using the other person for our own self-centered purposes more than we're actually connecting with them. And if they have healthy expectations of friendship, they'll quickly tire of us.

Disclosure in Established Friendships

We've seen the crucial role that disclosure plays in getting friendships off the ground, as well as how TMI and TLI can cause friendships to fizzle out before they start crackling. But you've avoided those misfires, and now you've got a roaring campfire going. You and your friend are happily roasting marshmallows together (I'm definitely taking this metaphor too far). Disclosure continues to play a vital role in maintaining and, if desired, deepening established friendships.

As friendships grow, reciprocating shallow disclosures, like how your commute was that day or what you thought of the PTA meeting, becomes relatively unimportant. What becomes crucial is reciprocating intimate or deep disclosures—as in, the real stuff of vulnerability, where we need support from our friends. As you get to know someone better, it becomes increasingly important to continue to reveal sensitive information to each other—that's how you become and stay close. Failure to do so can be hurtful and breed resentment, as it suggests you don't trust or depend on your friend.

That said, not every friendship needs to be deep to be meaningful. Anthropologist Robin Dunbar famously mapped our friendships into concentric circles: a few ride-or-die confidants at the center, and progressively larger groups of close, good, and casual friends radiating outward. Each layer plays a different role. Your best friends might drop everything to help you in a crisis. Your good friends might host your birthday dinner or check in after a tough week. Your casual friends might be your favorite people to sit next to at a wedding or the ones you grab drinks with a few times a year. None of these is "better" or "worse." They just offer different forms of companionship, each with their own rhythm and reward—and they thrive on different levels of disclosure.

In the closest relationships—friendships and romantic relationships, too, as we'll discuss in the next chapter—something remarkable happens: The need for strict, back-and-forth turn-taking in disclosure starts to fade. This isn't to say that openness disappears. Rather, it evolves. When your friend tells you something hard—say, they're worried about getting fired or they're stretched thin by caregiving—you don't feel obligated to chime in with your own hardship. And they're not expecting it. What replaces the rule of reciprocity here is something deeper: trust. The unspoken understanding that your turn will come. That when you need them, they'll be there.

Yet even in these intimate friendships, disclosure can go awry. Two common culprits? Misunderstanding and envy.

Tending the Fire: Handling Misunderstandings

Miscommunications are inevitable in friendship. That's not the problem. The real issue is what happens when we don't clarify, apologize, or explain how something made us feel. That's where friendships crack—or, if we're brave, where they deepen.

Consider Maria and Lena, who had been friends for over ten years, since their twenties. They'd had plenty of time to develop trust (shorthand: history). And still, a single text message nearly pulled them apart.

It was a bright sunny morning in April 2021. Things were looking up, Maria thought. The dark days of Covid seemed to be behind them; everyone was getting "vaxxed, waxed, and ready to party," as she and her husband kidded (they had a two-year-old little girl and didn't get out much anymore themselves). And, best of all for Maria, a month earlier, she and her husband had been overjoyed to find out she was pregnant with their second child.

The phone rang one day as they were running around in the kitchen, preparing breakfast while also tending to a newly walking toddler. Maria glanced at the number, and her heart raced. She had taken a blood test a week before to ascertain whether the baby had any genetic anomalies. Someone was calling from the hospital with the results.

"I'm happy to inform you that your baby is perfectly healthy," said the technician, whom Maria had quickly put on speaker. Her husband hurried over and held her hand. Perfectly healthy! The best words. "Would you like to know the sex?" the woman continued. "Yes," said Maria, as her husband squeezed her hand. He knew she really wanted a boy.

"It's a girl!" the voice pronounced cheerily. Maria gulped and mustered, "Oh, wonderful!" before putting the phone down and wiping away tears.

Her husband hugged her and said all the right things: "We're pregnant again! The baby is healthy! We're so lucky! The girls are going to be best friends!"

But nothing could console Maria in that moment, as guilty as she felt for being upset about something so minor. Having grown up with a younger brother she doted on, she had daydreamed about the same dynamic playing out in her house.

She dashed off a text to her good friend Lena, who was already in on the pregnancy: "The baby is healthy! But it's a girl. I know I'm really lucky, but I just can't help but feel really sad about this."

She paused briefly before hitting send. Lena also had one child, and Maria thought they wanted a second. Would this trivial disappointment upset her or seem a little insensitive? Maria quickly talked herself out of that concern. After all, when divulging the pregnancy several weeks before, she had asked Lena straight up whether it was okay to talk about the baby with her. Yes, Lena had said emphatically.

But Lena's quick response to Maria's text stung, with its unusual stiffness: "I'm sorry you're disappointed with your child's sex."

What followed made clear that, indeed, this was meant to be read as "You're so ungrateful and insensitive."

Maria immediately regretted having revealed her disappointment. What should have been a joyful moment had unleashed uncomfortable emotions and now a brewing conflict. She had hurt her friend's feelings. Badly. Maria felt terrible, but also annoyed. After all, Lena had insisted she didn't mind hearing talk about babies—that she welcomed it even.

Maria kept her irritation to herself. She kept everything to herself for the rest of the week. She and Lena stopped talking.

Fortunately, Lena reached out soon with an olive branch. "Let's have lunch!" That's when, over a tearful meal, Maria learned the full extent of Lena's efforts to have another baby. She and her husband had gone through several rounds of IVF. She described all the hormones. All the money spent. All the heartbreak. The waiting. The roller-coaster ride. And after all that, their doctor had recently advised them to consider adoption. There's so much to love about adoption, of course, but Lena desperately wanted to have another biological child.

"I feel terrible," Maria said. "My text about the baby was so insensitive."

Lena generously exclaimed, "But you didn't know! I kept it from you!" (It's worth noting, this wasn't a failure of first-order disclosure on Lena's part. She hadn't hidden her IVF struggles maliciously. She likely just hadn't been ready to open up, and that's okay.)

Disaster averted. One disclosure led to the problem, while another resolved it. All part of the same complex weave of reciprocity.

If she could turn back time, Maria definitely would not have sent Lena that insensitive text. But she's grateful that she and Lena talked about it. In the end, opening up about their underlying feelings deepened their friendship.

What saved them wasn't backtracking. It was second-order honesty. Maria let Lena see her guilt. And Lena didn't pretend she hadn't been hurt. She explained why she had been. That kind of mutual vulnerability doesn't erase the pain, but it can build something sturdier.

Having the courage to acknowledge disconnect and discomfort can shift tricky dynamics and misunderstandings. A log dropped the wrong way can smother a flame—or, with a little adjusting, catch a stronger spark.

A Particularly Thorny Challenge: Envy

Success can feel strangely isolating. In one study, a full 84 percent of people admitted to withholding a success from a close friend. Why? Because sharing our wins can feel awkward. Even our biggest cheerleaders are human, and humans have egos. That leaves the door wide open for envy. We often think the hardest thing to share is our pain. But often it's the good stuff.

Envy is what psychologists call an *upward comparison emotion*: It bubbles up when someone close to us gets something we want. And while it can be motivating (what researchers call *benign envy*), it can also get toxic fast (*malicious envy*). Malicious envy is so poisonous it doesn't just make you feel bad. It makes you want the other person to feel bad, too. And it's more likely to surface when someone's success seems undeserved—or when they share it in a way that feels arrogant.

So we might hold back from sharing good news with a friend to prioritize the friendship and avoid stirring up envy. Is that wise? The short answer: No. Because if and when your friend finds out about your success secondhand, they will be hurt—and so will the friendship. People feel insulted and less close to you when you hide your success.

Maybe we should try to soften the blow by humblebragging: Shoehorn the good news into a complaint or a self-deprecating aside. ("So annoying, I just got accepted into this fancy leadership retreat, and now I have to find childcare for three days . . .") But that tends to backfire. Humblebragging reads as insincere, and when people sense insincerity, they like and trust us less. They may even see us more negatively than if we'd just bragged outright.

How can you share successes in a way that elicits celebration rather than envy? First, consider the timing. Don't share your news when your friend seems low, irritable, or very busy. And think carefully about how to frame it: Can you reveal the struggle, not just the shine? In Alison Wood Brooks's research, people felt less malicious envy when they heard about someone's *failures* alongside their successes. Finally, don't be afraid to acknowledge your discomfort. For example, you might say, "I feel a little funny about telling you this, given what you've been going through, but I think it's important that we share our good news with each other." If envy is a poison in friendship, silence is what lets it spread.

A deeper fear: What will our success and their (possible) envy do to the friendship? This is especially a concern when the friend is someone we feel very similar to, which is the case for many close friends.

It's either fitting or wildly ironic, then, that one of my best friends is also a social scientist in my department at HBS: the aforementioned Alison Wood Brooks. Alison and I are about the same age, were hired around the same time, and became fast friends quickly, bonding over our love of ideas, our admiration for each other's work, and our healthy distrust of anyone who skipped the social part of social science. She's had great professional success. I've been fortunate as well, and yet I've felt envious of Alison's success from time to time. And yes, as you might have noticed a couple of paragraphs up, Alison studies envy (among countless other topics). Envy is hard to avoid in an environment like ours. At HBS, we're surrounded by brilliant, driven people, and nearly everything—teaching, publishing, research impact—is tracked, rated, or ranked in some way.

Expectations are sky-high, and comparisons are effortless. We love our work and feel lucky to do it, but it's a pressure cooker in the best and most intense sense.

Somehow, however, Alison and I had been friends for years before we finally broached the topic of envy. It took an opportunity falling right into our lap: A visiting scholar had just presented a paper in our lab meeting on the psychology of rivalry. He described the ingredients: similarity, competitiveness, repeated interactions. As he spoke, Alison and I started making eye contact across the room— silent, pointed glances that asked: *Is this . . . us? Are we rivals?*

On the walk back to our offices, she broke the silence. "Why *aren't* we rivals?" she asked, clearly meaning: How had we managed to skirt the pitfalls and become good friends? It was the perfect way in—curious, gentle, disarming. (There's a reason she's a world expert in the psychology of conversation!)

Once we were in the privacy of my office, I admitted, a little sheepishly, that I had sometimes felt envious of her. Of her confidence and her dazzling competence. Her speed. Her wit. Her EQ. Her gregariousness. Her ability to be both deeply warm and absurdly efficient. (Shall I go on?) And I told her the hardest part: that I felt petty for feeling that way. Because I love her like a sister. She didn't flinch. Instead, she nodded and smiled her megawatt smile, holding my gaze. Then, in her classic cool-as-a-cucumber, warm-as-a-sweater way, she said something like "Of course I've felt that way, too."

I think about that conversation often when a flicker of envy crops up. And every time, I'm reminded of how lucky I am to have a friend like her. Later that same week, she set me up on a date with a guy I was too chicken to ask out myself. Hardly the move of a rival.

Then there was the time she and another friend staged what can only be described as a real-world version of *The Bachelorette* for me.

They rented out part of a bar, invited a curated group of eligible men, and ran interference all evening—screening suitors, allowing only the best candidates to approach me. Years later, when I was in the throes of early motherhood, literally pulling my hair out while trying to breastfeed, Alison showed up at my house with the kind of exuberant positivity I didn't know I needed, a postpartum fairy godmother. (My own grand friendship gestures haven't been quite as well executed, like the time I sent her flowers with an anonymous note that read "To my love," thinking it'd be fun for her to guess. Turns out, leaving out just a *bit* too much information can briefly alarm even the most secure of husbands.)

Envy feeds on the belief that life is zero-sum: If you win, I lose. But my friendship with Alison has taught me that that's rarely the case. If someone you love wins, then you win, too. At the same time, it's okay to be thrilled for someone and also quietly envious of them. The work of friendship—the gift of it—is learning how to hold both. Disclosure can help us do that.

In Alison's case, I don't think it's a coincidence that she's so good at converting envy into generosity: She's an identical twin. She's had a lifetime of practice in seeing someone else's joy not as a threat but as part of a shared story.

When Words Fall Away

In the deepest friendships, something radical can happen: The need for disclosure recedes—not because we've stopped revealing, but because we've said enough. We've said the big things. We've survived the small things. We've used the reciprocity of disclosure to go beyond the need for closure all the time. We've built a rhythm of trust and presence that no longer relies on verbal volleying. Sometimes the most intimate form of reciprocity is simply sitting together in silence. This, too, is supported by research. Emma Templeton and colleagues found that while strangers often experience long silences as disconnection, friends experience them as moments of deepened connection—silences full of reflection and savoring.

In his memoir *We Should Not Be Friends*, Will Schwalbe tells the story of a friendship that began in one of the most jarring possible ways: mandatory, high-stakes self-revelation. As undergraduates at Yale, he and Maxey—a jock Will had initially avoided—were inducted into a secret society where, as part of the initiation, new members had to share their life stories. At his initiation, Will spoke about his fears and insecurities as a gay man. Afterward, Maxey approached and told him that while he didn't really "get" what it meant to be gay, he could relate to some of what Will had said: "You're as boy crazy as I am girl crazy." And somehow, that goofy line became a bridge.

Their friendship didn't blossom overnight. It waxed and waned over the years, shaped by health scares, relationship changes, and career ups and downs. But it endured. And in the final pages of the memoir, when both men are pushing sixty and facing serious health issues, Will describes one final scene: sitting with Maxey and realizing neither of them was talking—and neither of them needed to. "After almost four decades," he writes, "we had finally reached the point where we could talk to each other about almost everything. Perhaps the greatest reward for that was that we no longer felt compelled to say anything at all."

There's something quietly stunning about that moment. Not the absence of speech, but the presence of understanding. A kind of emotional warmth that no longer needs stoking. Some friendships are built on matches and kindling and carefully timed logs. And some, over time, become embers—slow-burning, silent, but still very much alive.

8

Finding Love

At thirty-five, things were going pretty darn well for me. It was the fall of 2016. My life was filled with friends and family. I was married and we owned a cozy home. I loved my job. Every morning I'd walk through the hallowed Harvard campus, over the footbridge that led to the business school where I work. As I took breaths of the crisp fall air, I'd often gaze below at the Charles River to see the sculling teams practicing. The sight of the paddles slicing through the serene water, and the sound of the rhythmic chanting of the rowers, always soothed my soul—which was good, because it needed soothing. I was grateful for everything I had, but my heart was in turmoil. My marriage wasn't working out.

Suffice it to say: After a lot of agonizing, we ended it. That's how I found myself dipping my toes into the world of online dating. Soon I became enmeshed in stressful games of strategy that involved seemingly endless disclosure dilemmas. Creating my profile on dating apps proved particularly complicated because of the question about my "relationship status."

"Separated," I first thought of saying. But didn't that imply my situation was messy or unresolved? As far as I was concerned, I was "Single" with a capital *S*. But was that misleading? How about "Divorced"? I mean, I'd already signed the papers. But what about the D-word baggage? Might as well say "loser at love," or so I ruminated in my more desperate moments. How to decide what to share!

Leaving the relationship status question blank would look like I was hiding something. I knew I wouldn't swipe right if that question weren't answered on someone else's profile. (Though I'm told that nowadays on these apps, there's much less emphasis on clearly stating relationship status.) So I continued to toggle back and forth between "divorced" and "single," and to deliberate the ethical and strategic implications of every other micro-reveal along the way. Beyond these questions of what to list for these basic pro forma attributes, there was the question of what to volunteer about myself in the freeform description. How much should I say about my personality traits, for example? How much about my hobbies? My profession? But, more to the point: Why, dear reader, should you care?

My personal dilemma highlights the many decisions we all face about when to open up. This is especially true when we're seeking a long-term relationship partner. Whether it's a dating app or good old-fashioned in-person encounters, we are constantly wondering about, and second-guessing, our decisions about how much to share, with whom, and when. It reminds me of the Miranda warning issued to those under arrest: "You have the right to remain silent. Anything you say can and will be used against you in a court of law." When you're tangled in the laws of courtship, it's the same.

Sometimes remaining "silent" isn't a realistic option, of course. One woman who uses a wheelchair showed the chair in her dating app photo. She had "really negative reactions of people either explicitly or implicitly saying the wheelchair was a no," she said, but her forthrightness paid off when she met the man she would later marry:

Not only was he not turned off, but he was eager to make sure that on their first date she would have everything she needed to feel comfortable.

A female friend of mine who is six foot four (yes, really) told me she put her accurate height in her profile to screen out men who didn't want to date someone taller than them. And in a reverse situation, when a newly single male friend of mine proudly showed me the dating profile he had created, I gave the requisite coos. I didn't say it, but I was especially impressed that he had listed his height—five-seven—plain and clear. And he's had no shortage of (great) dates. Being up-front is a helpful screener, as it saves a lot of time and confused looks and outright rejections. It can also signal that you are self-assured.

When you know that everything you share may be used against you, disclosure decisions can keep you up nights. But once you understand the complex dynamics of sharing or withholding when dating, when falling in love, and when sustaining a long-term relationship, I think you'll sleep better.

Swiping Right: When Less Is More

Whether on an online dating profile or a first date, there are lots of things about ourselves that we might consider revealing, beyond the basic facts of where we're from, where we live now, what kind of work we do, and our favorite hobbies. One of the most crucial things we want to convey is our personality—very simply, how it feels to be around us. Personality is key to chemistry.

The only way for someone to really find that out is to meet you. But if you're going the online route, people encounter your dating profile first, which should ideally provide a preview of what you're like and why someone might be interested in meeting you. So what kind of profile elicits interest from potential dates? Sure enough, scientists are interested in decoding this.

My colleague (and bestie!) Mike Norton and his coauthors examined whether the *amount* of information that daters supply in their online dating profiles affects others' interest in dating them. They compiled a list of 218 traits that online participants used to describe themselves, and then created random lists of these traits of varying lengths. Next they asked groups of online daters to read one of these lists of traits and indicate on a scale of 1 to 10 how much they liked the person being described. The daters also indicated if they had any traits in common with the person in the profile. What Mike and his team found was that participants preferred candidates who were described in fewer traits. That's right. *Less* information seemed to garner *more* interest. As the authors put it, prospects "who looked good from afar suddenly seem less attractive once more is known." Why exactly would this be?

Let's consider what people look for in long-term mates. Psychologists have found that similarity is a huge factor. When it comes to long-term romantic relationships, there's more evidence for the "birds of a feather flock together" hypothesis than the "opposites attract" hypothesis. Although opposites can be alluring for short-term relationships, when it's time to settle down, Eli Finkel and colleagues note, most of us want to be with people who are like us. That holds true not just for personality traits, attitudes, and beliefs, but also for physical attractiveness. We don't tend to go for the most attractive person, but rather for the person we judge to be about as attractive as we are.

One would think that the more traits a person listed in their profile, the more points of similarity the average dater would have with that person, increasing their interest in them. But no. Mike's team found that the longer a person's list, the *less* attractive others found them to be. While longer lists did indeed add up to more shared traits, they *also* resulted in more *dissimilar* traits. And the rub is that dissimilarity exerts a disproportionately strong influence on our judgments, especially on first impressions. In other words, dissimilarity is a bigger turnoff than similarity is a turn-on. Discovering that someone is different from you on a certain dimension creates a "dissimilarity cascade," which seems to escalate as you keep finding out about other things you *don't* have in common. Ironically, it's better to share less about yourself at the beginning. First a little mystery, then discover differences.

A related reason why "less (sharing) is more" in the initial phase of trying to attract a mate has to do with the well-established norm of reciprocity, which we learned about in chapter 7. Revealing a lot on your dating profile, or blurting out a bunch of things about yourself while flirting with a prospect at a bar, can be off-putting because you're not giving the other person a turn to reciprocate with sharing of their own—you're revealing out of turn. It would be like one person answering all of Arthur Aron's thirty-six questions before the other person did.

In the dating realm, sharing a lot right off the bat is off-putting for yet another reason. It can make you seem unselective. If you're so eager to open up to someone you barely know, it reads (accurately or not) as desperation. Studies have shown a curvilinear relationship between our perceptions of a potential mate's choosiness and our interest in dating them. That is, we tend to view people who are not choosy as unattractive, and the same goes for those who are so choosy that we conclude they'd never go for us. This latter point may sound counterintuitive.

Yet even early research on selectivity reveals a more complicated reality. In a 1971 study that would never pass ethics muster today, researchers hired a female sex worker to serve drinks to male patrons. She used one of two scripts, either playing it cool or coming on strong. Over the next month, men who got the cooler treatment called her more often to arrange "dates." The results suggest we're drawn to people who are moderately available—those who text back, but not immediately, or who hold back some information about themselves early on. No matter what kind of date you're looking for, awareness of these subtle dynamics can help ensure you intrigue rather than overwhelm.

On a date, revealing the "just right" amount of information about yourself (Goldilocks again) to signal self-respecting choosiness means making sure you're not monopolizing the conversation. You're asking plenty of questions about your date and striving to keep the sharing more or less symmetrical. Following this norm is trickier when you're trying to craft an online dating profile because there's no two-way flow of information. Yet this balance between sharing and showing curiosity can be achieved, Juliana Schroeder and Ayelet Fishbach found in their research. They culled and analyzed about two hundred profiles from two dating apps (Match.com and Coffee Meets Bagel), and (unsurprisingly) found that people most commonly wrote exclusively about themselves, divulging what they were looking for in a relationship, alongside other personal facts. This makes sense. A dating profile is, after all, a kind of self-advertisement. Interestingly, though, the profiles that consistently garnered the most interest were those that didn't focus *only* on self-revelation. These people projected a sense of curiosity onto potential date candidates, signaling a desire to get to know the other person through statements like "I'm excited to get to know you" and "I'd like to know more about your beliefs and thoughts." But engaging profiles like these were few and far between; when left to their own devices, only about 1 percent of love-seekers were this outwardly focused in their dating profiles.

Schroeder and Fishbach also found that when they gave this guidance prescriptively—when they advised people to express an interest in getting to know the other person in their profiles—doing so increased their appeal. This gesture is a powerful demonstration of the importance of reciprocity, as it shows that even in a noninteractive context, one in which mutual self-disclosure isn't possible, you can simulate its effect by signaling your interest in getting to know someone.

Knowing all this about how we present ourselves online should have made me a savvier dater. But as I found out, understanding the dynamics is one thing; navigating them yourself is something else entirely, even for someone who studies human behavior.

Every time I opened an app, I had to marshal my inner optimist and believe I could encounter a smart, attractive guy, and not just more dudes posing in front of a tiger or a Ferrari (or far worse). This seemingly built-in optimism in love-seeking is one more reason why revealing fewer traits leads to greater attraction. When we're highly motivated to find a mate, we're scarily good at seeing what we want to see in those early phases of dating. We don't know a lot about the other person yet, so we effortlessly fill in the blanks and assume the best, unless and until we learn otherwise (and even then, as anyone who has counseled a jilted friend can relate, it can take a long time to take off the rose-colored glasses).

This is also why, once you find someone who piques your interest in an online dating app, you should go ahead and meet them in person rather than continue to communicate electronically. People are really good at curating a particular image when communicating digitally, which means you're not actually getting to know them better, and vice versa. In fact, a long delay in meeting may actually *decrease* your odds of clicking in person. The longer you communicate virtually, carefully curating your respective images, the more time and data your brains have to optimistically "fill in the gaps" about each other. When you finally meet in person, you may both be disappointed by how reality has eclipsed the elaborate fantasy you built up in the preceding weeks or months.

That said, love is built, at least at the beginning, on idealization. Our initial fantasies are destined to go unfulfilled once we meet the people who inspired them, but if we're lucky, what we find is close enough to the fantasy that we are not so disillusioned we walk away. In any case, overoptimism is adaptive. It motivates us to soldier on, dating in the face of repeated failure. As Mike Norton and his colleagues summarized, we keep trying because we sense "the lone payoff may be worth the many disappointments."

I know those disappointments very well (and the eventual payoff, too). A year after starting my right-swiping journey, I was still at it. I'd get home from work and mix myself a stiff vodka soda—a drink chosen not for taste but for its exceptional booze-to-calorie efficiency. I then forced myself to spend thirty minutes on a dating app. Yes, it was a chore.

And then there were the dates, like the time I bailed out of a moving vehicle (no, I wasn't in danger—just pathologically averse to the looming goodbye kiss I didn't want to dodge by, you know, using my words). There was the impromptu TED Talk I received over drinks about 3D-printed self-tying knot devices, which, I was assured, were a major breakthrough in modern engineering. There was the steady parade of "not as advertised" profile pics. Call me callous, but eventually, before first dates, I started pausing outside the bar to peer in. And if I didn't recognize the guy from his photos, I'd just . . . keep walking. I felt like I learned more about human behavior in a couple of months of dating than during my whole PhD.

That's an exaggeration, but I did see how dating distills our hopes, biases, and insecurities into one messy cocktail.

Don't get me wrong. I met many wonderful, interesting, smart, talented, attractive people, but none of them felt right. One evening, I finished my daily chore with a right swipe—I always liked to leave on a positive note. The warm smile and soulful eyes grabbed me. His profile, otherwise fairly opaque, told me that he was educated, shy, and a former professional soccer player (swoon!). The dearth of information only contributed to the allure. Instead of his name, there was just a letter: "C." Well played.

The next morning: Boom! A match! I was so excited. *And* he had messaged me. The adrenaline was pumping. After a bit of witty, flirty banter on the app, we decided to meet up. I arrived first at the designated location, a coffee shop in Harvard Square. Shortly thereafter, a tall, devastatingly handsome, well-dressed (but not overdressed) man strode up to me.

"Hi, I'm Colin," he said. I vividly remember the exhilarating, flushed feeling I had the moment I met him—it just felt so right. In the past, I had rolled my eyes at the idea of "love at first sight." Now, here I was, immediately and seriously smitten.

Since I'd only just met the guy, I knew better than to gush. But it wouldn't be me if I didn't overthink it just a little, if only privately.

Was I already a disclosure nerd by then? Absolutely. I knew the studies, the pitfalls, the magic of gradual intimacy. But theories are one thing, and sweaty-palmed reality is another. In real life, you don't think, "Ah yes, I am now executing a stage-appropriate self-disclosure." You just feel your heart racing and wonder if you're saying too much—or not enough.

Long before I met Colin, I had already spent plenty of time mulling over all kinds of disclosure dilemmas I wished someone had studied. What's the best way to let someone down—be honest, tell a white lie, split the difference, or just ghost them altogether? Should you admit to feeling nervous before a first date, or keep your cool at all costs? And how exactly do you pull off "moderate availability" in real life without seeming either disinterested or desperate? I even started experimenting informally: swapping out different profile pictures to see which ones got the best response, and recruiting a close friend to ruthlessly edit my profile.

Since those dating days, behavioral science has actually started to catch up to some of these questions—especially when it comes to ghosting. Research led by Coral Zheng at Cambridge University suggests that ghosting hurts more than direct rejection. As Coral told me, "Almost any type of response, no matter if it's very brief and non-substantive or a polite rejection, is better than ghosting." Her studies showed that ghosted people often question themselves more harshly, feel sadder, and recover more slowly than those who receive even a quick, simple rejection.

These findings confirmed my hunches at the time. As an online dater, I kept a spreadsheet (yes, really) tracking how I let people down and how they reacted: whether I got angry texts, if they tried to recontact me and, if so, how persistent they were. Over time, I learned that telling some version of the truth ("You're wonderful, but not the right match for me") usually landed the best, though sometimes it did provoke angry replies or painfully self-punishing follow-up questions ("Was it something I said?" "Can you tell me what I did wrong?").

Ghosting, on the other hand, often triggered unrelenting waves of unanswered texts—which, frankly, made me feel bad, too. And when the tables were turned, I learned firsthand just how brutal ghosting could be. It's the ambiguity that gets you. You find yourself inventing increasingly elaborate explanations. Maybe they're just super busy? Maybe they lost their phone? Maybe they're trapped in a remote jungle with no Wi-Fi? Anything but the most obvious, most painful answer: They're just not interested in you. In those moments, it's incredible how easily we forget Occam's razor—that the simplest explanation is usually the right one. When I was clearly, unambiguously rejected, it hurt, but it also spared me from endless rumination and from sending a slow trickle of increasingly mortifying unanswered texts to the ghoster.

Still, I'll never forget the time I made a major misstep. On an early date with someone, our conversation drifted to the highs and lows of online dating (because nothing says chemistry like trading bad Tinder stories), and I foolishly laid out my "rejection strategy"—explaining the exact line I usually used to wrap things up early on when I realized the spark just wasn't there. And then, just a couple of weeks later, when I realized we weren't right for each other, I kid you not, I completely forgot and used that same script on him. It was only as I hugged him goodbye and launched into my usual well-intentioned line—"You're wonderful, but . . ."—that I realized, mid-sentence, that I was quoting myself. Word for word. From our earlier conversation. I slinked away in shame.

After enough bumps and bruises, I developed a first-date strategy of my own. To keep dates at least mildly entertaining, I used to play a little game. I would ask the person question after question and wait to see if they reciprocated by asking me a question. Some did, some didn't. One time I lost count after asking twenty-three questions and receiving none in return. This scheme kept me somewhat amused, but it was also a test. Would they show an interest in me by asking questions, or were they so self-involved or socially inept or insecure that they would only share about themselves?

Back at the Harvard Square coffee shop, I was floored by Colin's performance. He was beating me at my own game. I found myself revealing as much, if not more, about myself than he revealed about himself. I even stopped keeping count of my questions! Could this be happening—had I found my "lone payoff" that Mike and his colleagues promised, the one that would finally make all these disappointments worth it?

Getting to Know You: When Less Is Too Little

Excited as I was about Colin, I continued to date (as did he, I assumed). But my dates with other guys felt different, underwhelming even when they weren't bad. For one reason or another, even though some were certainly interesting and attractive men, all of them were no-gos.

Meanwhile, I couldn't get Colin off my mind. I was particularly nervous for our second in-person date, which didn't happen until six weeks after our first, thanks to our travel schedules. In the meantime, we'd exchanged the occasional flirty text, but by the time the second date finally arrived, I was feeling both excited and anxious. I was excited because the restaurant he chose was very close to his apartment. But I was also nervous because he didn't know that I was separated (hadn't revealed that in my profile!). In hindsight, my concern was justified. He later admitted that if he'd seen "separated" or "divorced" on my profile, he probably wouldn't have swiped right.

This brings us back to the question of how much to share about yourself in your dating profile. For immediately apparent qualities that people, rightly or wrongly, have strong opinions about—like height—I think it's best to be up-front in our dating profiles, as in the case of my six-four female friend (and my five-seven male friend, for that matter). But for an invisible feature that might have a social stigma attached, like marital status, it can behoove you to hide it at first, so they get to know the real you without their misconceptions or stereotypes getting in the way. But, as a divorced person, I'm biased! And clearly, these decisions of what to reveal—or to not reveal—are as much art as science. You have to do what feels right to you, then see what happens.

On date two, after a few sips of wine, I worked up the nerve and mustered something like "There's something I've been wanting to tell you about myself . . . I've been married before. And I'm in the process of finalizing the divorce."

I felt incredibly relieved—and Colin didn't seem disturbed by my revelation. Still, I could feel my face getting hot. A waiter drifted over, sensing an opportunity to check in, but Colin caught his eye and gave a quick, almost imperceptible shake of his head. The waiter vanished. Colin turned back to me, calm and steady, as if I hadn't just dropped a small emotional bomb on the table.

Perhaps because I had confided that I had misrepresented myself, he went on to tell me about an omission of his own. It was also a big one, but of a very different kind. On our first date, he had told me that he had no siblings. Now he revealed that he had had a brother who had died under tragic circumstances at age twenty-three, when Colin was in college. Colin had spent a lot of time in therapy to learn to cope with the loss. The experience had shaped who he was, he said. Wow, I thought, what a disclosure. But it didn't feel like TMI; on the contrary, it was a moment of connection because he was showing his willingness to be vulnerable with me.

After dinner ended, I lingered outside the restaurant, wanting so much for him to kiss me. I know, I could have initiated. Call me old-fashioned, but I like the guy to make the first move. That said, I was giving off every possible signal short of puckering my lips and making come-hither motions with my index finger. Nothing! So I requested an Uber. Then, just as it drove up, Colin reached toward me, and my neck instinctively popped back Scarlett O'Hara–style. Yes, I needed kissing, badly. And Colin sure knew how. Now when we walk past the corner where that kiss took place, I stand under the gaslit lamppost and we reenact the moment. It never gets old.

Shortly after that second date, I revealed something that was so TMI that with anyone else it might well have sunk the relationship. At the time, I was going through IVF to freeze my eggs. One day when I was having trouble finding someone who could pick me up from my egg retrieval appointment (they don't let you just grab an Uber after sedation like that—believe me, I tried), I went out on a limb (way out!) and asked Colin if he could do it. We had shared enough with each other even that early in our relationship that I sensed it would be okay. I was right. He happily obliged, as if this was the most normal request ever—something we still laugh about.

In the weeks and months that followed, we continued to have ever deeper conversations, fueled by intimate self-disclosure and an increasing sense of comfort and interdependence. We had entered the period in a relationship when my earlier advice not to share too much becomes irrelevant.

One revelation from the thirty-six questions process described in chapter 7 is that, especially when building relationships, most of us, at some level, really *want* to share about ourselves. Yet we often refrain from asking probing personal questions on dates for fear of seeming intrusive or "too interested." After all, overasking can be its own delicate hazard.

In fact, we tend to overestimate the risks of sharing and underestimate the benefits. In one study, I asked participants to imagine revealing increasingly personal information to someone they'd just started dating—from something mild like "I had a dog growing up" to something more vulnerable like "I sometimes struggle with depression." They expected that revealing more sensitive information would push their date away. But a separate group—who imagined being on the receiving end—actually liked the discloser more, not less. The bigger the reveal, the bigger the gap between what people feared and how others responded. Which helps explain why things heated up between me and Colin after we both shared uncomfortable truths on our second date.

Understandably, people are often reluctant to share intimate information early on. But as my friend Leilani discovered, sometimes it's a risk you have to take if you hope the relationship is going to develop into something meaningful. Leilani lives with borderline personality disorder, which, as she describes it, "means that my emotions are mercurial, I flit between avoidant and anxious attachment styles, and so feeling secure in a relationship is probably the hardest thing in the world for me." So when she started dating Jordi, she kept emotional distance despite being very interested in him. Leilani could tell that Jordi was getting increasingly confused about her standoffish behavior. Fortunately, she recognized that she was sending the wrong signals and knew that if she didn't take the risk of opening up about why getting close is so hard for her, she might lose the opportunity for a meaningful relationship.

Like so many of us who want so badly to share something important about ourselves and yet hesitate at the moment of truth, she found herself caught in a kind of disclosure deadlock. On numerous occasions, she tried to explain her behavior but stopped short, scared of how he would react. "I danced around the topic of my diagnosis, like saying it was admitting something was wrong with me," she says. She ached to be completely honest, but felt literally, physically frozen in place, as if she were being choked.

Finally, she figured out how to push through: Focus on a single word at a time. "If I take a deep enough breath, if I can make my lungs move and feel the air circulating around my heart, then I can say one word," she told me. "And if I can say one word, I can say a sentence, and then suddenly I'm talking."

Indeed, sometimes the hardest part of a difficult conversation is simply initiating it. As one of my favorite Peloton instructors, Robin Arzón, says: "You're here. You already did the hardest thing." The same goes for difficult conversations. Just starting is at least half the battle.

And so, one day, as Leilani and Jordi sat outside enjoying a quiet moment together, she found the words she needed. "Jordi, do you know anything about borderline personality disorder?" And just like that, Leilani's fear was gone—even before Jordi had said anything.

"I see you, I get it, you're safe," he finally said, simply. "I admire you so much."

Leilani's fear is now completely gone. "He understands—he sees where I'm coming from, and knows how to help me," she told me. "Next time we talk and I pull back, he knows how to come get me."

Yes, Leilani was fortunate that Jordi responded so well. There was a real risk of rejection. Her revelation could have scared some suitors away. That said, it was a necessary risk for Leilani to take; if she hadn't, her behavior might have alienated Jordi and prevented them from ever having the chance to get closer to each other. Similarly, while divorce can be stigmatizing, it was something I really *needed* to tell Colin. It had become (for me) the elephant in the room, something blocking any deepening of our relationship. And like Jordi, Colin responded in exactly the way the science would predict: He let my disclosure deepen his understanding of who I was—someone who, like him, had been through hard things and learned from them.

We had crossed one hurdle and come out stronger. But a few

months later, I found myself staring down a much bigger one: Should I be the first to say I love you?

The Matzoh Ball

In a *Seinfeld* episode, George tells Jerry that things are going well with his new girlfriend, Siena—so well that he's thinking about "making a big move." He's debating whether to say "I love you" and, as always, turns to Jerry for advice. But Jerry warns him that expressing love without being sure of hearing it back is risky—"a pretty big matzoh ball hanging out there," as he puts it.

If you've ever been tied up in knots about telling a partner you're in love with them, you are not alone. And we have good reason to fret about it. As herd animals, we're extremely sensitive to social rejection. After all, for our early ancestors, being separated from the herd was "often equivalent to death," write psychologists Geoff MacDonald and Mark Leary. Unrequited love is about the strongest, and most painful, form of social rejection there is. Our brains respond to both falling in love and losing love in ways that are surprisingly similar to addiction and withdrawal. Roxy Music was right: "Love is the drug."

It's no wonder that people are often very nervous to be the first to say those three words, which can so easily fall on ears that don't want to hear them or on lips that don't know what to say back. Sometimes the other person simply doesn't have strong feelings for you. Or they don't have such feelings *yet*, and your premature love declaration might prevent them from developing them. Or so you worry.

But what counts as premature, and how can we tell whether an "I love you" will (adversely) affect the future of a relationship? To answer such questions rationally, we would—theoretically—like to see an experiment where we randomize people who are falling in love to either say "I love you" or not, and see whether saying it increases or decreases closeness. Obviously, we can't do that study. But scientific literature does provide some comforting hints at an answer; it suggests that saying "I love you" is unlikely to discourage your partner from falling in love with you *if* you have a reasonable basis for saying it—that is, because you've gotten to know each other fairly well, have become increasingly interdependent, and are, well, super into each other. If you say it before then, you risk coming across as over-eager, possibly desperate. But if you say it to a partner who is already headed toward love, it might actually accelerate that process. Studies have shown that daters commonly feel *flattered* if you tell them you're into them, and even like you more as a result—especially if the interest feels genuine and selective.

Now let's change sides. What should you do when you're on the receiving end of a love profession, but you're not quite there yourself?

Consider my friend Emily, who was surprised when Julius, a guy she'd been dating for just a few weeks, sprang the big matzoh ball on her one night as they were canoodling on the couch.

"I can't stop thinking about you," he said. "I . . . I think I love you!"

Caught off guard, Emily giggled nervously. Her mind raced: *What do I say?*

She and Julius were both in their mid-thirties and, as their online dating profiles disclosed, both looking for a long-term partner. Emily didn't want to mess around; she wouldn't have been dating Julius if she hadn't seen him as a potential lifelong match. So she wasn't quite as flummoxed by his seemingly premature declaration of love as she would have been when she was younger and dating more casually. Plus, they were getting along great, seeing each other often, and she had no desire to date anyone else.

What, then, kept her from adding her own matzoh ball to the soup?

What was slightly odd to Emily was that Julius blurted out his feelings before they had slept together. They had great chemistry but were taking things slowly. With his sexual restraint, Julius seemed to be signaling that this wasn't just a fling, and she appreciated that. In fact, he was already making plans for them to spend a weekend together at a cozy log cabin. He had hinted that he hoped they would wait until then to be intimate. But can you be secure about the long-term potential of a growing love before you've really explored the physical side of the relationship? Emily wasn't sure.

She wasn't ready to respond with a similarly unguarded profession of love. But she didn't want to discourage Julius, either. So she said what felt true, if on the optimistic side, couching it in a slight hedge: "I think I might be falling in love with you, too." Emphasis on "think," which he had said, too, and "falling," which was a bit more tentative than actually *being* in love, yet still reciprocal-ish. Tentative as Emily's words were, Julius seemed thrilled by them, and they kept the relationship going.

The romance ramped up after that. The next month, they went on their getaway, which was as romantic as it was passionate. And soon, in keeping with their competitive natures, they also had a favorite quip: One of them would say, "I love you," and the other would respond, "I love you more." Fifteen years and two kids later, they still say it to each other.

What about our friend George Costanza? Well, true to tragicomic form, things don't turn out quite as George hopes. When he finally gets up the nerve to tell Siena how he feels, she simply replies, "I'm hungry; let's get something to eat." Crushed, George resigns himself to move on, only to later discover that Siena is deaf in her left ear, so he begins to ruminate about whether she had actually heard him. "Oh my god," he says to Jerry. "She probably never heard it. Don't you see what this means? It's like the whole thing never happened."

Hoping that maybe she *does* love him, he tells her again, this time being sure to say it clearly into her *right* ear. But as any *Seinfeld* fan would intuit, it just was not meant to be for George: "Yeah, I know. I heard you the first time," she says politely but firmly.

Siena's response was painful in the short run. But in the end, by having the courage to persist and ask again, George at least obtained useful information. There was no ambiguity. She just wasn't that into him. And though this is really hard to hear, if it were you, wouldn't you want to know sooner rather than later? The value of closure in the form of a clear, unambiguous "no" can be a great gift in the long run.

So, again: When is the right time to say "I love you"? Social psychologist Joshua Ackerman and his colleagues identified one critical factor to consider: whether you've been intimate yet—that is, whether you've had sex. (The paper focused on heterosexual relationships only. Science is still playing catch-up when it comes to studying the romantic lives of LGBTQ+ individuals.) As it turns out, Emily's intuition was spot-on. Science suggests that pre-sex professions of love can be a little suspect. While women may say "I love you" when they genuinely feel it, hetero men may be more apt to have something else in mind.

If you had to guess whether men or women are the first to say "I love you" in heterosexual relationships, which would you pick? When researchers posed this question to passersby on a university campus, about 64 percent guessed women. This jibes with the widespread beliefs that women are more emotionally expressive and relationship-focused—beliefs grounded in truth. If women have more of these qualities, the argument goes, then they'd fall in love faster, and therefore be quicker to express it.

But Ackerman's studies revealed that it was actually men who were more likely to profess love early on. This was not necessarily because they were in love, but, in many cases, because it was a strategy to initiate sex. Overall, people focused on short-term relationships were especially likely to make early declarations of love, often before they'd had sex. This pattern held for both men and women, but it was more common among men. From an evolutionary perspective, this makes a certain sense. Because women bear the higher reproductive costs of sex (I'll say!), men may have evolved to use emotional expressions—like saying "I love you"—as a tactic to signal commitment and increase their chances of mating.

The takeaway: People should pay attention to when "I love you" enters the conversation. If it comes unusually early, it might be more about seduction than sincere attachment. And as for whether it works? That depends on the goal.

Now, obviously *you're* not wondering whether saying "I love you" will lead to sex. But hypothetically, if someone *were* wondering, we don't have hard data to answer this question (imagine that experiment!). Still, relevant research suggests that early love declarations can sometimes be effective in this regard—especially if the listener is open to something casual or, I hate to say it, craving connection so badly they can't quite see it for what it is. (Been there.)

But as for the deeper question, if you're looking for a real, lasting relationship, what's more likely to move that along: saying "I love you" before sex, or after? It's tricky, because the act of saying "I love you"— even early—can be powerful in its own right. Expressing deep affection can make the other person feel chosen, wanted, special—and sometimes that spark can jump-start genuine love, even if the timing isn't perfect. But on the whole, the available science suggests that waiting tends to work better. Professing love after sex is more likely to be perceived as sincere and more likely to strengthen real attachment.

Of course, skepticism isn't always warranted. Recall that Julius had made a pre-sex profession of love to Emily. He had suggested consummating their relationship at a log cabin getaway that was over a month away. To Emily, that wait felt like an eternity in the throes of infatuation, and so it made Julius's words seem sincere rather than strategic.

It's worth noting that different cultures approach this dilemma very differently—or sometimes not at all. In a classic *60 Minutes* segment, Morley Safer spoke with people in Finland about saying "I love you." One of them, a female American expat reporter, balked at the very idea of Finns being so forthcoming: "No, oh my God, no, no. Not . . . even lovers." And a Finnish man, with the air of someone who clearly felt he was being openhearted—even a little indulgent—offered: "Well, I'd say you could say it once in a lifetime. Say, you have been married for twenty years; perhaps your spouse is on her deathbed. You could comfort her with saying 'I love you.'" When Safer laughed, he was swiftly corrected by the Finnish man: "It's not funny." Even a famous Finnish tango singer admitted she found it easier to say "I love you" in English than in Finnish, joking that Americans "love almost everybody." When a Finnish man says "I love you," she added, "he really means it." Different cultures set very different expectations for what emotional expressions mean— and when, if ever, they're appropriate.

As for Colin, me, and the L-word that stood between us, a couple of months in, we were still dating strong and had even taken a trip together—a romantic getaway to Palm Springs. I wanted to tell him that I was starting to fall in love with him and that I wanted to date exclusively, but I was scared of being rejected. So I tried to side-step my way into the good ole relationship definition conversation.

On our last day in Palm Springs, as we lounged in bed (ah, those glorious prechildren days!), I asked, "So, what are we?" in a tone feigning nonchalance—my "cool girl" voice (as Colin now affection-ately slash mockingly calls it).

You might think such a conversation would lead us to express our clear affection for each other. You'd be wrong. Colin said, "I'm hav-ing fun, and I like how things are." Womp womp.

Fighting back tears, I rolled over and opened Tinder.

The Course of True Love Never Did Run Smooth

From that moment on, our relationship started to stagnate. Colin had been much more reserved than I from the start, and up to this point, that hadn't bothered me. But I became increasingly upset that he wasn't sharing more about himself. I started to believe not only that he had a hard time expressing how he felt but that he was emo-tionally unavailable.

I've come to appreciate that attachment styles are extremely rele-vant to expressing feelings in relationships, especially love. But it wasn't just our innate attachment styles that prevented us from saying the L-word. Securely attached or not, it requires courage to say it, be-cause the possibility of rejection is real. It's hard to risk putting your-self in such a vulnerable position, even if the potential for reward is great. And what I didn't appreciate at the time was that I had as much trouble sharing my true feelings as Colin did. I could reveal all sorts of embarrassing things I'd said and done—something of a passion of mine, as you may have noticed. But when it came to sharing my deep-est concerns and feelings, I was at least as buttoned up as he was, even though I somehow thought he was holding back more.

Colin was the one who finally worked up the nerve to be the first to (tentatively) declare his love. A couple of months later, one night he guardedly announced, "I think I love you."

My response? "Me you, too."

Me you, too?!

My nervous retort might have blown the door open for a full-throttled discussion of how much we'd come to care for each other. We might have laughed and shared our great relief that the truth was out. Instead, we both flinched. Rather than seeing our messy, imperfect confessions as a beginning, we each mistook the other's awkwardness for ambivalence—and quietly pulled back.

In the days that followed, we avoided talking about our feelings. Our conversations moved in the wrong direction, becoming more and more superficial. For my part, I was sure Colin regretted bringing up the L-word. After all, he'd only said he *thought* he loved me. I decided he'd realized he didn't after all. We wound up in relationship limbo, neither of us willing to open ourselves up further to the other.

A few months later, we broke up. Or rather, I broke up with him. I didn't want to. But I didn't know how to breach the wall that had been put up between us. I could have just said that. Instead, I plunked myself down beside him on my pleather couch one morning right before I left for work.

"I think we should break up," I said stoically, without emotion.

"Okay," he said, and sat there with a blank look on his face. After an awkward pause, he said, "I'll take my stuff when I leave today."

His apparent indifference was extremely hurtful, yet also perversely validating: *I guess he really didn't love me after all*, I concluded. As I walked over the footbridge that day, I beat myself up. I was devastated. Why hadn't I opened up more? I was sure he was the love of my life. How could I have just let him walk away? Why hadn't I worked harder to keep him? To get close to him? I had missed out on a wonderful man.

I kept thinking about Colin over the next couple of months, full of regret. Then, one lovely fall day, I came home to discover an envelope on my doorstep. "Leslie K. John" was written on it in Colin's distinctive hand. No address . . . and no stamp. He had been here, himself, in the flesh!

The letter was so Colin. In it, he finally declared his love.

"If I'm being honest with myself, I know that I do love you," he wrote. "I know I was scared of admitting that I was in love with you, scared to embrace how I was feeling about you, and scared to let myself be in love with you and all that meant." He also hedged, in his usual deeply reflective way, saying that he was still sorting out what was important to him in life. "I know that these are questions that I need to answer for myself . . . questions that I may never have a clear answer for, something that I wrestle with daily."

I waited a couple of days. In hindsight, I don't know why I waited; being coy by not responding immediately seems cruel. The time for being "moderately available" had long passed. He had sent me such a clear message. He had put himself out there. When I did finally text him—"Hiya . . . got your lovely letter"—my heart was pounding. I stared at the screen expectantly. Thirty seconds passed. Then a minute. Then . . . those three dots appeared. He was somewhere, looking at his phone, and thinking of me! We were communicating! We were back in touch! I was ecstatic. We quickly made plans to go for a walk Sunday afternoon.

The walk turned into a whole-day affair, and day turned into overnight. We fully declared our love. I confessed that I'd felt he was "the one" for me the very first time we met, but that I had been scared he wouldn't feel the same. I didn't want to seem desperate. He said that he had felt the Exact. Same. Way.

It all just came out. And it felt *so* good.

I am so lucky that Colin had the courage to write the letter. A year later, we were married. I got my second shot at love. And "me you, too" has become our favorite expression of endearment.

Keep Talking

Opening up to each other doesn't stop at "I love you" or even after marriage. Relationships often falter not because of dramatic betrayals, but because we stop nurturing deep, mutual understanding. While most long-term couples invest in getting to know each other early on, many eventually abandon that active discovery—despite the fact that we're always changing. Daily routines take over: work, kids, dinner, TV, bed. And the space for meaningful conversation quietly disappears.

But, ironically, another cause is that *we come to believe we know everything about our partner.* As a result, we stop working to get to know them better. And if they do the same—asking fewer questions, showing less curiosity, not listening—we stop sharing about ourselves as well. This can lead to two people who once felt intensely close to each other gradually drifting apart. Writer and podcaster Cat Sims vividly described how she and her husband became distanced from each other in this way: "There was no cheating, drama or violence. We had simply turned our backs on each other and kept walking until we couldn't see each other or shout across the distance." Some people stay in these empty relationships, together but alone. A 2018 survey of married couples age forty-five or older found that almost 33 percent of those questioned said they felt lonely.

I saw this dynamic unfold firsthand in my own relationship.

Early on, Colin and I had been cautious about expressing the intensity of our attraction—and that was probably wise. After all, we didn't yet know each other, and sharing too much too soon can backfire. But as time went on, our early reserve hardened into a habit of withholding. We assumed we understood each other, and we stopped sharing feelings as openly. And sure enough, it led to misunderstandings, distance, and emotional stagnation.

It turns out that one of the biggest dangers to ongoing connection in relationships isn't outright betrayal or growing apart, it's falling into the same trap we did: what researchers call *mind-reading expectations*.

When we're in a close relationship, many of us expect, often without realizing it, that our partner should be able to read our mind: when we're upset, why we're upset, even whether we want to talk about it. But no matter how close two people feel, it's surprisingly hard to intuit what someone else is thinking.

In one study, researchers videotaped 158 heterosexual couples as they discussed an area of conflict in their relationship. Afterward, the partners each separately watched the recording of the discussion. Every ninety seconds, the video was paused and the participants were asked to write down what they had been thinking and feeling at that moment, and to guess their partner's thoughts and feelings. Partners accurately inferred their partner's thoughts and feelings *only about one in five times* (and no, women were not significantly more accurate than men).

On one hand, this might seem impressive when you consider the infinite number of things a person could be thinking at any given time (as a supreme mind-wanderer myself, this feat is not lost on me). But on the other, it's shockingly low given that these couples had been together for twelve years, on average, and were guessing about a conversation they had just had.

Many studies have assessed *empathic accuracy*, the "ability to accurately infer the specific content of another person's thoughts and feelings." For neutral, everyday conversations, empathic accuracy hovers at around 30–35 percent. For more emotional or conflictual conversations, it's lower, around 20 percent. Worse, most people, and especially married couples, overestimate empathic accuracy. TL;DR: We think we know what's on our partner's mind more than we actually do. And that gap between what we think we know and what we actually know is where trouble starts.

The mistaken belief that our partner can read our mind is surprisingly common—and surprisingly harmful. It's linked to a host of relationship problems, including the tendency to undershare. One study found that people with higher mind-reading expectations were less likely to communicate their needs clearly, believing their partners should "just know." Unsurprisingly, those expectations were associated with poorer communication and lower relationship satisfaction. (And when researchers developed a scale to measure these beliefs, they didn't mince words: They called it "a measure of dysfunctional relationship beliefs.")

Believing your partner should be able to read your mind inhibits open communication. You don't say what you need because you believe, if only implicitly, that you shouldn't have to. In turn, when your partner inevitably fails to meet your (unspoken) needs, resentment builds. This leads to passive-aggressive behavior, misinterpretation, and a vicious cycle of disconnection.

Where do these mind-reading expectations come from? They're built from a few powerful illusions. First, romanticism: the belief that true love should grant supernatural understanding. No words necessary. Poetic, but obviously wrong. Second, overconfidence: the persistent belief that we're great at reading each other's minds, when in fact research shows that our confidence has little to do with actual accuracy. Third, the curse of familiarity: the belief that, over time, we've learned our partner's personality so well, we think we know everything else, too. But knowing someone's personality traits— their stable tendencies—is not the same as knowing their shifting thoughts and feelings in a given moment.

Researchers who have studied relationship satisfaction in couples see a connection between marital happiness and a feeling of being heard and understood by one's partner. The more regularly partners opened up to each other, the more they reported feeling understood, which in turn predicted relationship well-being.

Suggestive evidence of the power of this regular sharing has been found in a number of correlational studies that asked partners to keep track, every day, of how much they revealed facts (such as "I got my performance review today"), thoughts ("I think my boss might not understand how hard I work"), and feelings ("I feel so frustrated because I sense I'm underappreciated at work"). Some studies also asked partners the extent to which they revealed positive emotions versus negative emotions. In all of this research, disclosure was positively related to relationship satisfaction. And the most powerful type of disclosure was about one's feelings, both positive and negative.

The point is simple, but easy to forget: Staying close means staying curious. Ask and listen.

Once I understood this, it was like a fog lifted. I realized that so many of the little tiffs Colin and I had weren't about deep incompatibilities. They were about silent, dangerous assumptions. I would be upset that he hadn't responded to an unspoken need. He would be hurt and confused, not knowing what he'd done wrong. Now, when Colin seems unresponsive, I remind myself: It's not that he doesn't care. It's that he can't read my mind. And when I tell him what's bothering me, it's a game-changer.

———◇———

Over time, Colin and I have learned that what feels like over-communicating is often just communicating. What feels like oversharing is often just sharing. And that, I think, is the real labor of long-term love. Not the grand gestures or perfect timing, but the small, stubborn work of staying known to each other.

In the early stages of dating, a little mystery is essential. But once you build a life with someone, mystery isn't what keeps the connection alive—understanding is.

The night Colin and I got married, we stayed up until four in the morning, still talking. It felt like the perfect start—and the right reminder—that love isn't just about finding someone you understand. It's choosing to keep meeting them for a date, over and over again.

9

Workplace Revealing (and Concealing)

t was August 2010. On the cusp of completing my PhD, I was on the academic job interview circuit. For many people, including myself, this process feels like an intimidating, sometimes mortifying, hazing ritual. Universities hoping to recruit send senior faculty delegates to a giant nerd jamboree (aka academic conference). Job candidates hoping to *be* recruited are interviewed, one by one, in hotel rooms. Most of the schools use standard rooms, though the fancier ones get the suites. Either way, it's odd knowing your future hangs in the balance while gazing at generic hotel paintings on the walls.

The interview itself usually consists of some awkward small talk to start (academics are not typically the most socially skilled of creatures), followed by the candidate's forty-minute research presentation. It's decidedly low-tech—the candidates move from room to room, clutching little flip books of their printed research presentations. Besides feeling like a product you have to sell to prospective buyers, the process can also sometimes feel kind of sketchy. For example, in one interview, I walked into the hotel room to find two male professors lounging on hotel beds. They were fully clothed but weren't wearing shoes. It was a strangely intimate scene for a professional interview, and one that, in retrospect, really captured how off-kilter the whole setup was. I mean, when's the last time you saw your boss in their socks, not to mention lounging on a bed? But I sat down at the table beside the beds and proceeded to give my presentation, as if it was the most normal thing ever. I could barely concentrate, wondering if they were trying to unsettle me. Were they testing me? Had they already decided against me, so they figured why not get some relaxation time in?

There was one interview that I was particularly excited about. It was with a very fancy university (they had the grandest suite I'd ever been in), and I saw the interview as the first step toward landing my dream job. A lot was riding on it. I happened to know from credible intel that I was on this university's short list, so I felt like the job was mine to lose.

I desperately wanted to make a good impression, and as you might imagine, this did not help matters. Instead it ramped up my nerves, and when I get really nervous, I sometimes find it hard to preserve the professional decorum expected in such circumstances.

At 10:00 a.m. sharp, heart a-pounding, I pressed the suite's doorbell and was warmly greeted by one of the senior faculty members. I made my way to the single lonely chair set up in front of the phalanx of my interrogators: eight prestigious professors, all of them men. I sensed they could tell I was nervous. I desperately hoped I wouldn't start visibly sweating.

Now, at this point, I should mention that I have an unusual résumé. I alluded to it earlier, but before I went into academia, I was a professionally trained ballet dancer, a career I abandoned due to chronic injury. One of the faculty members gallantly attempted to break the tension with some lighthearted small talk. Glancing at my résumé, he said cheerily: "I see you were a ballet dancer. I was, too!"

I read this as a friendly, self-deprecating joke, because the professor didn't have a dancer's physique. He was obviously seeking to calm me down. I should have just chuckled, appreciatively acknowledging how thoughtful he'd been.

Instead, I cocked my head quizzically, looked him up and down, and said in a sassily sarcastic tone: "Clearly."

What had I done?! I had just openly insulted one of the people I was trying to ingratiate myself to. Way to go, LJ. I had revealed my snarky side in a context in which I was expected to be buttoned up and professional. And in just one word, no less!

The room went dead quiet, and my face burned red. The professor I'd just inadvertently insulted stared at me in a stunned stupor. Another professor—someone I knew fairly well from a research collaboration—leapt up and asked if I wanted a glass of water. An act of mercy, offering me a chance to pause and reset. But in a stupor of my own, I delivered blurt number two: a curt "No, thanks." (He brought me the water anyway, because he is a better person than I am.) Still, in the thick silence that followed, I swear I caught a faint smile from him, like he was rooting for me. Maybe because he'd already seen me at my most vulnerable. Lucky him—he'd been there for the infamous pee story. And I'd lived to tell the tale. Maybe that story hadn't tanked my budding reputation after all. Maybe, just maybe, it had helped. But now I'd landed myself in another blurtatious mess. Damn it.

The rest of the interview was a bit of a blur. I couldn't get out of the room fast enough. Poof! In one thoughtless moment, there went my chances at clinching the professorship I'd worked so hard for. I said the obligatory thank-yous and fled.

The workplace presents a distinctive set of disclosure dilemmas, beginning with the strange fan dance of interviewing. We are trying to put our best foot forward, to convince our potential employer we're a perfect fit and consummate professional, yet we're asked "What are your weaknesses?" and "What are the biggest mistakes you've made?" Even the seemingly laid-back "So, tell me about yourself" can feel like a trap. Where should we start? We don't want to come across as a paid political ad for ourselves, as so rehearsed that our interviewer gets no compelling sense of who we really are. Maybe we should show some of our fun side so we don't seem uptight. After all, who wants to work with an automaton? What outside-of-work interests might we share? Or will that make us seem self-involved? How do we find a sweet spot of revealing?

In the workplace, there are strong norms about how we "ought" to present ourselves, *especially* in first-round interviews: confident, upbeat, professional. In other words, pleasant, if a little dull. The work self we project is often quite different from our rest-of-life self—which, in my case, can be a little snarky. By inadvertently revealing my sarcastic side in that interview, I had violated this strong norm.

There has been a lot of buzz in recent years about the benefits of "bringing your whole self" to work. There's some evidence for those benefits. Letting others see more of you than you might ordinarily show them forges bonds, including in the workplace. We saw this in the early pandemic, when hardened leaders suddenly turned into endearing softies the moment their toddlers mischievously ran into their home offices. But for compartmentalizers who prefer to keep work and personal life separate, the "bring your whole self to work" movement can be something of a nightmare. For others, like me, it's freeing. But this new terrain is filled with land mines, and it can be hard to know when you're going to step on one, as I felt I did in that interview.

The question of how much of our authentic selves to share at work is a pivotal one. It's also a difficult one to answer. We want to share enough to feel understood and connected to others, but not so much that we alienate people or cause them to question our competence or our seriousness.

Making matters even more complicated, each workplace has its own culture and its own norms about the degree of self-disclosure that's deemed appropriate. That doesn't mean they're clearly articulated; usually far from it. We have to discover them. And by no means should everyone decide to simply conform to those norms; bucking them might be good not only for one's own happiness and engagement at work, but for the whole team and for society at large.

So how do we find the right balance? What are the trade-offs between being a little more open at work and keeping strict professional boundaries intact? How much "backstage access" can we give to our colleagues and our bosses without risking our workplace image?

Backstage versus Front Stage
(and Vulnerability versus Transparency)

Our Canuck friend Erving Goffman from chapter 2 considered this matter at length in his pioneering work on self-presentation in daily life. His conclusion? Very little of the true self should make it to what he called the "front stage," where we should carefully control our "performance"—that is, what we say and do to shape others' impressions of us. But times have changed since the mid-century Goffman era, when a buttoned-up persona was standard workplace fare.

According to my colleague Monique Burns Thompson, who works closely with members of Gen Z, "Today's generation craves a level of openness that is different from when I was a young professional."

NYU organizational scientist Julianna Pillemer's research suggests that revealing aspects of our backstage selves at work, when done thoughtfully, can help us build rapport and stand out in a good way. In workplace contexts, she recommends what I'd call *discerning authenticity*—a balancing act that involves giving colleagues some, but not total, access to our inner lives (Goldilocks, again). When done well, Pillemer argues, it helps build trust and sparks more meaningful conversations. Over time, this kind of thoughtful openness can deepen workplace relationships, enhance collaboration, and even improve performance.

What does it mean to be discerningly authentic—to be open in a thoughtful way? Pillemer specifies two types of backstage access. The first, which she calls transparency, involves "conveying openness" by giving people a window into your thoughts, beliefs, or preferences. For example, you might say, "I've always been more drawn to the creative side of things, even though I'm technically in a data-heavy role." This kind of sharing can carry some risk—especially if your perspective is unpopular or unexpected—but it generally offers only a glimpse beneath the surface.

The second level of access, which Pillemer calls vulnerability, goes deeper and carries more risk. It involves "sharing potentially sensitive inner states such as intimate emotions," especially negative ones—like admitting that you feel insecure about public speaking or disclosing a disability that might lead others to underestimate you. For instance, someone might say, "I get nervous presenting in front of senior leadership, even when I know the material cold" (revealing a performance-related insecurity), or "This kind of ambiguity is tough for me. I like having more structure, and I'm trying to get more comfortable with the gray area" (revealing a trait that might not align with organizational norms). One shortcut I find helpful is to think of transparency as *cognitive* openness and vulnerability as *emotional* openness.

In contexts where impressions really matter, the line between transparency and vulnerability becomes a strategic one. Pillemer doesn't draw a hard line, but she emphasizes that vulnerability is riskier—especially in high-stakes, evaluative settings like job interviews, where disclosing insecurities might chip away at perceptions of competence. If in doubt, transparency is the safer bet. Vulnerability should generally be avoided in those contexts unless, say, it's framed as a story of growth or overcoming a challenge ("I used to struggle with public speaking, so I joined Toastmasters").

Even when you're explicitly invited to share something personal—like in the dreaded "tell me about a weakness" question—transparency often does the trick. You might offer cognitive openness: "I think better in writing than I do speaking off the cuff." You could also frame it as growth: "I've learned to prep more deliberately for meetings so I can articulate my ideas clearly in real time. But if you give me a moment to organize my thoughts, I'll always bring sharper insight."

This kind of thoughtful disclosure lines up with what Pillemer would call transparency: revealing how your mind works in a way that's candid but not risky. Vulnerability, by contrast, might involve admitting that you often doubt your abilities or fear being judged—disclosures that could raise red flags unless carefully framed.

Still, even in high-stakes settings, being a bit more open can help. One project, led by organizational scientists Celia Moore and Dan Cable, looked at what distinguished job candidates who actually received offers. Among the most-qualified applicants, those who aimed to present themselves authentically were more likely to be seen as "real"—and those seen as real were more likely to be hired.

Being seen as real didn't mean baring your soul. It often meant offering a realistic, unpolished description of how your mind works. For example, a candidate might say, "I usually approach a problem by asking what success would look like in the end, and then reverse engineer from there. That's just how my brain tends to work." That's transparency—not vulnerability. It reveals a personal cognitive style but doesn't expose weakness or invite judgment.

And it wasn't just *what* people said that mattered—it was how they said it. Moore and Cable found that candidates who were perceived as authentic didn't just share more personal content; they also spoke differently. Their language had a more fluid structure and included more function words—those little glue words like "what," "then," "from," "just." Use too many of these types of words and you might start to sound verbose and inarticulate. But when used in moderation, they're what make speech feel natural and conversational. And this tone—more than polish—was linked to better hiring outcomes.

To see the difference, take that same transparent self-description and strip out the function words: "I usually approach problems by defining success, then reverse engineer that." The meaning's the same, but the tone shifts. The second version sounds more mechanical, less like someone thinking aloud and more like someone checking a box.

Authenticity may also act as a kind of screening mechanism, helping us match into jobs we're more likely to love. In one study, people who approached the interview process more authentically—by offering a more realistic picture of themselves rather than a polished facade—ended up feeling more satisfied and committed on the job. The researchers suggest this happens because being real helps both sides better assess that elusive thing we call fit.

So, that fancy job interview of mine? It was with Harvard Business School. And, contrary to my expectations, I got the job! I daresay that I did so not *despite* my transparency, but in part *because* of it. In fact, the colleague I blithely insulted became a close mentor of mine. To this day, we get a kick out of regaling others with our "origin story." As he puts it: "When you insulted me like that, we all thought, 'Hey! She's a jerk, too! She'll really fit in here and will do great with our [very demanding!] students.'" At that point in time, the dominant classroom schtick was what I would call the "jerk" archetype, as exemplified by a colleague of mine who, if a student came in late, would freeze like a statue, glaring at the student until they sat down. Only then would he restart class—and from the very beginning, to really amp up the shaming. I don't act as jerky as that (at least, I hope not), but I do think that my snarky side has served me well at HBS.

Evolutionary biologists might label such candor in interviews as a "costly signal"—an honest but risky display of one's true value that is aimed at distinguishing oneself from less qualified poseurs. In the animal kingdom, to evade death, weak animals are motivated to mislead predators into thinking that they are stronger than they are, just as potential mates want to present themselves as more desirable than they actually are. For humans, this can mean posting puffed-up résumés or exaggeratedly good-looking photos on social media. Up to 96 percent of candidates misrepresent themselves during job interviews in various ways, including concealing aspects of their background and "embellishing" their qualifications. This leaves those who are the "real deal" with a challenge: how to signal that they're not like the poseurs. They have to show predators, potential mates, and employers alike that they're formidable opponents, reliable partners, and excellent hires.

In job interviews, offering a bit of backstage access can do more than help assess fit. It can also signal fitness. That is, it shows you're confident enough to take a risk by being real. Indeed, as I later learned from my HBS colleagues, when I stuck my neck out in my interview, I conveyed strength—if only unwittingly. You could say that the person who has the balls to insult a tenured Harvard Business School professor is, by definition, confident—a necessary trait to survive teaching here.

I should also say that although I am confident in my abilities, I also had a massive case of imposter syndrome. This I did not reveal in my interview, for good reason. Such a revelation is getting into *vulnerability* territory, which, as we're about to see, is much riskier—especially for those who face revealing a hidden stigmatized identity or invisible disability.

Revealing Invisible Stigma at Work

When organizations tell workers to be themselves, they often ignore the reality of how hard it can be for young people, for minorities, for LGBTQ individuals, and for women at large to develop credibility in the workplace.

—ALICIA MENENDEZ, MSNBC NEWS ANCHOR AND AUTHOR

The riskier level of the backstage access that Pillemer discusses—revealing some kind of vulnerability—is what Alicia Menendez is talking about. People with mental illness, neurodivergence, or nonheteronormative sexual identities may experience stigma, but their identity is not one that will inevitably "out" itself unless they choose to disclose. This leaves them with a distinct dilemma: Do they reveal their truth and risk bias, or do they conceal it indefinitely and deal with the emotional and cognitive burden of doing so?

Take an executive education student of mine, whom I'll call Karen. I'm not proud to say it, but I found her particularly, well, relentless in conversation. But the usual nonverbal tricks I use to stealthily get someone to stop talking—reducing eye contact, or moving closer to them, which makes it awkward for them to keep talking—didn't work on her. She was unresponsive to my social cues. Finally, I resorted to inelegant tactics like cutting her off and not calling on her. Ugh.

So I was surprised when, after the final session, Karen gave me a big hug and thanked me. I was astounded that she appeared to actually like me.

What she said next surprised me further.

"I have always had a hard time in social contexts, because I am neurodivergent," Karen said, a tad too close to my face. That explains *a lot*, I thought to myself. And I think she may have been telling me this because of the way I had treated her in class—that she understood why I had acted this way and felt the need to explain herself. Suddenly it started to make sense. She went on to say how she found the abrasive approach (my words, not hers) I had taken with her to actually be *helpful*, precisely because it was clear and direct.

"My condition can be an advantage sometimes," she continued. "I do a lot of high-stakes negotiations, and I don't get easily offended by others. And it's easy for me to be direct, which can be a huge asset in these contexts. I work in a male-dominant country but have no problem negotiating very assertively with senior businessmen. But at the same time, I really struggle to pick up on social cues."

Wow. If I didn't feel like a jerk before, I certainly felt like one now. I found myself wishing that she had told me about this aspect of herself at the outset of the course. Without knowing anything about her, I had dismissed her as a problem I had to deal with.

Upon reflection, I understand why she hadn't said anything. Her approach is consistent with the advice of organizational scholar Chloe Kovacheff, who studies disclosure of stigmatized identities. As Kovacheff told me, "It's best to reveal invisible, potentially stigmatizing identities only when you have established trust with the person." This is a hard-to-resolve dilemma, because on the one hand it means that, as Kovacheff put it, "the onus is often unfortunately on the individual to speak up and to craft accommodations." On the other hand, this means that even if the person or organization you are dealing with is willing to help you, they can't, because they have been left in the dark about your needs. All this has made me wonder how I might create an environment in which people feel comfortable speaking up to me sooner.

Some historically stigmatized identities, like sexual orientation, are not inherently tied to job performance but may shape workplace dynamics in ways that are often overlooked (e.g., by bringing valuable unique perspectives). But they can be dangerous to reveal, as they can also give rise to discrimination. Some people don't feel compelled to reveal these identities, while others do. As we've seen, actively hiding such an identity generally isn't good for your well-being. And if the truth comes out anyway, you risk losing the opportunity to reveal it on your own terms. Whether, as well as when, you choose to reveal depends largely on whom you'd be revealing to. Disclosing to your workplace friend would likely be safer than revealing to a manager, especially if it is someone you've heard make insensitive remarks. Once you've worked there for a while and your manager has seen and commented on your value to the company, you might consider risking disclosure. But if you think silence is the better option, then that's what you should choose.

In general, research shows that avoidance approaches to workplace disclosure can undermine well-being. This includes not revealing an important aspect of yourself, or revealing it but then downplaying or inhibiting it. For example, an LGBTQ+ employee working in a conservative office might feel pressured to conceal their identity, avoiding any mention of their partner or changing pronouns when talking about their personal life. Over time, this kind of self-censorship can be taxing (like keeping a secret). The mental strain of filtering conversations, dodging personal questions, or even fabricating details to fit in can erode a person's sense of belonging, making it difficult to fully engage with colleagues or feel invested in the workplace.

As another example, a person with ADHD may choose to downplay their diagnosis by avoiding requests for accommodations, even if those accommodations could enhance their productivity. Unlike concealing one's sexual identity, which may be about preserving social acceptance, withholding an ADHD diagnosis might stem from a fear of being seen as less capable or needing special treatment. Yet, in both cases, the act of hiding comes at a cost. Some identities and conditions can remain hidden indefinitely, but others—like needing workplace accommodations—eventually lead to an unavoidable conversation. This raises a crucial distinction: Some disclosures are optional, while others are inevitable.

From Hidden to Visible:
When the Reveal Is Inevitable

There are some aspects of identity or circumstance that eventually become visible, no matter how much we might want to keep them private. This might be the case for someone transitioning to a different gender, for example. Pregnancy and progressive disabilities also fall into this category. Similarly, needing workplace accommodations to perform one's job can lead to an unavoidable disclosure even if the underlying condition remains invisible. In these cases, the question then is not whether to disclose but when, to whom, and how.

I think a good example of that is my friend Christine's experience. Christine doesn't remember seeing the stars as a child. She does remember *trying* to see them, on a Grand Canyon family camping trip. She recalls "straining to see what everyone was talking about, convincing myself that I could see some twinkle as well." Other than glimmering stars at night, Christine could see just fine. So for years after that trip, she didn't think there was anything wrong with her sight. But looking back, she realizes, "I couldn't see the stars because my vision was already deteriorating."

Christine is a former colleague of mine, and she has retinitis pigmentosa, a condition that has been causing her to gradually lose her vision. Eventually she will lose her sight entirely. As of now, she has night blindness, and during the day, her sight is restricted to a small, circular field right in front of her. She has no peripheral vision.

When she interviewed at Harvard ten or so years ago, her eyesight was better, though she was legally blind at the time. I wouldn't say she *hid* her blindness during the interview; I think she downplayed it. And I think that was a shrewd call on her part. If I had known the full extent of her impairment, I might well have jumped to the conclusion, if only implicitly, that the obstacles she faced would compromise her ability to do the job. And I would have been wrong, for she's a prolific researcher and an excellent instructor.

So it's not surprising that Christine often chooses not to disclose her vision loss. "I worry that others will think less of me," she says. "And sometimes it's just not worth getting into." She has developed a lot of work-arounds. For example, she might bump into a pole at the playground while with her kids and laugh it off, saying to the other parents, "Goodness, I'm clumsy today." Or she might ask a delivery person for help signing for a package by saying, "This font size is too small for me to read without my glasses, could you please point out where I should sign?" If someone shows her photos on their phone of their children (guilty as charged!), she might exclaim about how adorable they are, even though, as she puts it, "For all I know, they could be showing me a photo of a potato."

Notably, it was only after Christine's job offer was signed, sealed, and delivered that she met with the head of our unit as well as HR to discuss the accommodations she would need. Christine's approach jibes with Kovacheff's recent research on this topic. "If you're risk averse, then it's safest to wait until *after* you get the job to disclose your invisible disability," Kovacheff told me. But she also aptly noted that just as job interviews give your prospective employer insight into you, "they are also an opportunity for you to get to know your employer—to see if *you* like *them*. So from this perspective, if you broach the topic of disabilities, you can see how they respond. If they don't respond well, you can be pretty sure it's not a good environment for you." Indeed, I liken job interviews to dating, in the sense that bad behavior, or even more subtle red flags, at the outset are highly diagnostic.

But once you've gotten the job, Kovacheff advises being proactive. "You don't want to wait for your employer to learn secondhand," she says. Moreover, her research suggests pairing the disclosure of the disability with a request for whatever specific accommodations you will need. She finds that when you handle the issue this way, managers tend to appreciate your candor and view you as both competent and confident about your own worth. If you choose not to disclose your disability and it becomes apparent anyway, it may come across as something you're trying to hide. That perception can negatively affect how others see you.

I didn't begin to recognize the true severity of Christine's vision loss until a year or so after she had joined our unit. It was a bright summer evening and I was leaving work. As I walked across the foyer's checkered marble floors, I looked out through the regal over-sized doors. Just beyond them, at the top of the small set of outdoor stairs, I saw a figure shuffling in place. She was feeling gingerly for the edge of the step so she wouldn't fall. I realized it was Christine, and what she was dealing with on a day-to-day level dawned on me for the first time. That brief glimpse made me begin to realize just how much accommodation she likely needed to do her job. And as someone who is fascinated by disclosure dilemmas, I had all kinds of questions for her about how she decided what and what not to reveal, and to whom.

When I caught up with Christine, she mentioned how her disclosure dilemma had evolved. Her sight has worsened such that she benefits from a cane, but is loath to use it because it deprives her of the ability to hide her disability, and therefore to insulate herself from the stigma associated with it.

As she told me about the latest stage in her vision loss, she paused. The fact that she still faces the dilemma—whether to use the cane or not, to reveal or conceal—felt, in a way, like a privilege. It meant she could still walk on her own. She is going through an unimaginably difficult transition: a loss that unfolds in slow motion, a disclosure dilemma that isn't resolved by choice, but by inevitability.

Christine's experience shows how slow and complex the process of disclosure can be. Timing and control can matter as much as the choice to share. This kind of struggle isn't unique to disabilities. Many people face a similar but distinct challenge when it comes to pregnancy. Like Christine's condition, you can conceal pregnancy for a while, but eventually your body announces it for you. And as with disability, pregnancy can trigger biased assumptions about competence. Studies show that people—especially those who don't know you well—often assume a pregnant employee won't perform at their peak. That stereotype can lead pregnant employees to overcompensate by working harder than ever. I've heard accounts from female profs whose teaching ratings tanked during their pregnancies.

I was fortunate enough to basically avoid the teaching-while-visibly-pregnant issue during my first pregnancy. That's because I entered my second trimester in March 2020, just as the pandemic was shutting the world down. Teaching moved entirely online, an incredible piece of luck for me (though it was hugely disruptive for most). I got to teach the entire course *sitting down*. (In a fun flourish at the end of the course, I stood up and revealed my big belly for the first time, to audible gasps!)

A professional acquaintance, Tracy, avoided any discussion of her pregnancy while leading a months-long, in-person course. She wasn't visibly pregnant at the outset, but very much was by the end. The whole time, however, she acted as if there was "nothing to see here." Even when participants approached her to congratulate her, she'd dodge the topic. "I didn't want to be defined by my pregnancy," she explained. "I didn't want all the baby goo-goo, ga-ga cooing. I wanted the focus to be on the material, not on my belly." I get it (though I wouldn't have been able to pull this off).

But Tracy *was* forthcoming with the leadership at her institution, wanting to be proactive in preparing for her leave and for any support, like a backup instructor, that might be required. Her boss was all business. "He did what he had to do, but he didn't seem exactly excited for me," she recalls. She was disappointed that he couldn't at least pretend to be happy for her, but all things considered she was fortunate.

One qualitative study that tracked managers' reactions to pregnancy disclosure found that while 61 percent responded positively, expressing excitement and support, a significant percentage couldn't bring themselves to express even the most basic of social niceties.

One woman reported that the first thing out of her supervisor's mouth was "You can't bring the baby into work with you," while another recounted that her boss said that "he legally can't be anything but happy for my husband and [me]." "Legally" he couldn't, because there are now some legal protections for pregnant people—but they don't stand in the way of the kinds of prejudices so many women encounter on the job.

As with other invisible but consequential identities, navigating pregnancy disclosure at work requires careful judgment. Purposefulness is key—not just disclosing for the sake of it, but identifying the person who is most relevant and positioned to support you. If your supervisor seems likely to respond poorly, consider starting with someone else—ideally, a trusted senior colleague who can help you strategize or even facilitate the conversation. There's no single right approach, but being deliberate about timing and audience can give you more control over how the news is received and what follows from it.

What They Need to Know
(and What They Don't)

Craig was a leader at a large Fortune 100 company. He had been employed there for almost twenty years and had worked his way up the corporate ladder. But his demonstrable successes belied some turmoil in his personal life. His performance suffered. Then one day, his boss called him into her office to tell him his position had been "eliminated," as they say. He froze as the news was delivered and then finally managed to get up and stumble out of the room.

Over the next few days, Craig thought about things. He realized he needed help. He opened up to friends, family members, and mentors about what he and his family had been going through. His mentors seemed to grasp that Craig's family struggles had been directly contributing to his flagging performance. "I think there was a beginning understanding that if I had had cancer, they would have carved out space for me to deal with that and then come back to work in a more full-time, committed way," he says.

Craig's revelations led to an acknowledgment that there was a "systemic problem" in the company that he might be "uniquely suited to address," he reflects. His company specially designed an executive role for him: Mental Health Ambassador—a cause he felt passionate about. Craig has become an advocate for mental health awareness and support in the workplace. He's coached hundreds of employees around the world on how to seek support inside their organizations.

Of course, it was easier for him to be vulnerable after he had been laid off, because he felt he had nothing to lose. Looking back, he wishes he had opened up sooner. It would have saved himself, his family, and his organization a lot of hardship.

So I was surprised when, in an interview, Craig told me, "When people come to me and say, 'Should I tell my story?' I always say no."

He quickly clarified, and this clarification points to a key lesson from chapter 5: Know your purpose. Many people feel an urge to share their struggles at work, but they often haven't paused to ask *why*. Without a clear purpose, it's harder to make good disclosure decisions. So Craig's default "no" isn't a rejection of disclosure, but a call to be deliberate. As we'll see in chapter 10, there *is* a time and place to share your story in the workplace. Doing so can spur great positive change. And we'll hear from Craig again there.

But first, I pressed him on a common workplace dilemma: What if you're struggling with a mental health issue that's affecting your ability to work? In such cases, some level of disclosure may be necessary—to request accommodations or to take medical leave. Craig, having navigated these challenges firsthand, emphasized that this kind of disclosure doesn't have to be overly detailed. "Often, people feel the need to explain everything to their boss," he said. But that can be risky. Instead, he offered this reminder: "Under US law, you don't have to tell anybody anything. You're entitled to medical leave without sharing your diagnosis." Whether you're requesting leave or seeking smaller day-to-day accommodations, Craig advises: "Only share what you think we need to know in order to support you."

He also stressed the importance of context. "It's vital to be aware of your environment. What's the culture of your company? Is it open and supportive? Or more secretive and skeptical? What's your relationship with your manager?"

"I'm an alien from another planet—what specifically should I look for?" I asked.

He chuckled. "Look for signs that mental health is taken seriously. Does your company acknowledge Mental Health Awareness Month? Do they offer employee resource groups for mental or physical health? If all you see is yoga, it might not be a very safe culture." He added that your boss may not be the right person to tell: "If you sense they won't be supportive, go to HR instead."

Finally, he gave a wise reminder: If you start a conversation and it suddenly doesn't feel right, you can always pause. Maybe your would-be confidant is having a bad day. Or maybe they're just not who you thought they were. Either way, trust your instincts. You can always come back to it later. Because once it's out, you can't put the genie back in the bottle.

How Crying Became Taboo, and How to Do It Well

Showing our emotions in the workplace also involves risk. We face pressure to conform to norms of professionalism: Offer service with a smile, pretend the customer is always right, respond agreeably to criticism about our performance that we know is unfair. It's all rather exhausting, isn't it? Despite our best efforts to stay stoic, there are moments when emotion simply takes over. Whether we're in a cubicle, at the front desk, or standing at a podium, we sometimes can't hold back the tidal wave that comes during difficult times. When that happens, we know we've overstepped a line, and that people will likely view us as less-than.

But one type of taboo emotional outpouring that particularly interests me is crying. And that's because I've always excelled at it. When I was a kid, I was picked for the Crybaby role in the National Ballet of Canada's production of *The Nutcracker*—the little girl at the opening Christmas party who marches to front center stage and has a crying fit when she doesn't get what she wants. I gave a rousing performance.

Even now, as a grown adult, I'm still an avid crier. I cried tears of joy when my son pooped in the potty for the first time, and tears of frustration when my kids Just. Won't. Stop. Whining. I've also literally cried over spilled milk—my own, which I had just pumped, in the middle of night, in a brief stint away from my firstborn.

As you might already suspect, my crying bouts are not restricted to my personal life. I have cried at work several times. Once was in front of my stiff-upper-lip, *very* British boss. He stared at me, bewildered, and handed me his bespoke embroidered silk handkerchief. The thing is, for the life of me I can't remember what I was even crying about that time! Another time, in an academic seminar, I was moved to tears by the beauty of two almost completely nonoverlapping bell curves (yes, I'm a giant nerd). And once, while teaching a class of ninety students, I broke down while recounting the story of the *Challenger* space shuttle disaster.

Crying at work is common. One survey of 3,078 employees found that 83 percent reported they'd done so. Nonetheless, I've always felt shame about my loss of emotional control. That is, until one of my outbursts gave me a new perspective on crying at work.

I was giving a seminar at a prestigious university to an audience of fellow academics. Now, the thing about this university is that it had a reputation for heckling guest speakers—a reputation that it deserved (fellow B-school profs: Yes, it was the school you're thinking of). Knowing this, I had prepped like crazy, trying to think of every possible unfair question and various ways of handling them. But nothing could have prepared me for what was to come.

Almost immediately after I began speaking, the audience started whispering and murmuring. Then they began peppering me with antagonistic questions and interrupting me without even raising their hands. I kept my composure well enough for quite a while, but then the dam burst. I started crying uncontrollably. Ugly crying. Not subtle in any way, shape, or form. But valiantly (or foolheartedly?) I soldiered on . . . getting a few words out here and there before being overwhelmed again by those involuntary, shuddering cry breaths that make crying in public feel especially undignified.

I was crushed, sure my reputation and thus my career was ruined. Chiding myself to regain my composure, I suddenly felt emboldened. After a few minutes, the ugly crying had subsided, and I almost regained the ability to talk like a normal human. (For the record, what follows is the cleaned-up version of what I said; IRL I was still sniffling like a malfunctioning accordion.)

"I want to stop the talk for a moment," I said, "and tell you why

I am crying. It's not because you are asking challenging scientific questions. I welcome such questions, as it helps make the science better. I'm crying because of your tone. You're being rude. You've interrupted me repeatedly. You're being belligerent. That may seem like acceptable and appropriate, and even 'normal,' behavior to you—because this is how you act for every speaker. Your reputation for being uncollegial precedes you. As an outsider, I'm telling you that this is not normal. And it's not okay. It has to change. And not just because it's hurtful. But more importantly, because it gets in the way of us having a good exchange of ideas."

When I had finished, the room was silent. Everyone had listened attentively. As I completed my talk, they were on their best behavior. After the talk, the meanest of the meanie profs came up to me and apologized. A few weeks later, I was heartened to hear from a colleague who had just given a talk there. She told me the same group had been extremely collegial to her. Maybe I had sparked change, if only temporarily.

I didn't know it at the time, but I had stumbled onto a strategy that researchers Elizabeth Baily Wolf and colleagues have found makes crying in the work sphere more accepted, and even meaningful. As expected, they found that people perceive employees who cry at work—both males and females—as less competent than those who do not. But there was a caveat. Workers who attributed their tears to being emotionally invested in their work were deemed more competent than those who said they were simply feeling emotional.

In one experiment, for instance, participants read about an employee who, upset about the dynamic in their team, breaks down crying in front of their teammates. For some participants, the description of the worker ended there. Other participants went on to read that the worker then said either, "I'm sorry, I am just really passionate about this," or "I'm sorry, I'm just really emotional about this," or "I'm sorry." Participants rated the workers who chalked up their tears to passion for their work as more competent than the other criers. By linking tears to a commitment to one's job, we may be able to mitigate negative reactions to our crying.

Wolf's result was the same regardless of whether the crying worker was male or female. But the finding may be particularly helpful for women, because they are more likely than men to face backlash for crying at work. This stems from common stereotypes that men are emotionally stronger than women. As a result, men's tears are often assumed to be triggered only by something truly serious. So they get a pass for crying, because they're assumed to cry only for cause. Whereas for women, "the weaker sex," our tears are interpreted as a sign of weakness and overreaction.

Yet plenty of men do cry at work, and often for the same reasons that women cry. One study of 1,500 US office workers on performance reviews found that 25 percent of the male respondents said they had cried in response to a critical review whereas 18 percent of the women reported the same. Nonetheless, it is only women who are stigmatized by crying. As Olga Khazan concluded in *The Atlantic*, "crying joins the list of things—makeup, raising kids full-time—that people look down on simply because women do them more."

As the fascinating history of crying lays bare, the warped notion that women cry more and should therefore be considered weaker is a social construction—one that has undergone dramatic change over time. According to writer Sandra Newman, male crying has been viewed positively for most of recorded human history. For centuries, men cried openly without fear of being stigmatized as weak or feminine. Newman notes that in Homer's *Iliad*, "the entire Greek army bursts into unanimous tears no less than three times." Accounts of men weeping are found throughout historical records of the Middle Ages, not only in the West but in Japan as well. Medieval romances are full of lovelorn knights crying because they miss their ladyloves. And the Bible is chock-full of tears—from the "weeping and gnashing of teeth" found throughout, to the shortest verse in the King James Version, "Jesus wept." In the eighteenth century, it was very common for upper-class men to cry in their courts; as Tom Lutz put it in *Crying: A Natural and Cultural History of Tears*, men were "viewed as brutes" if they did not cry.

So what happened? How did we get away from a point where "grown men actually forced themselves to cry publicly in the hope of impressing their peers," in Newman's words? There isn't a written record to tell us what caused male tears to dry up. But Newman and others posit that the rise of urbanization and industrialization led crying to become verboten for men. During the Middle Ages, most people lived in small villages surrounded by kin, an environment conducive to intimacy and free-flowing emotions. With the rise of cities, more and more people lived among strangers. Men entered factories and offices, where emotional expression was discouraged as a time waster. They were trained to suppress their emotions so as not to interfere with "the smooth running of things," writes Lutz. In that, they had a long head start on women, but the pressure on women not to cry once they also entered these workplaces became intense.

I hope that we become more appreciative that crying—even on the job—can be a healthy and justified way to release difficult emotions. Research has found that crying is good for us in many ways. It relieves stress and releases oxytocin and endorphins, which ease both physical and emotional pain, and it can activate the parasympathetic nervous system, which helps regulate our emotion by slowing down our heart rate and breathing. As we'll see soon, even leaders have been known to set a healthy example by crying from time to time.

10

Why Great Leaders Share More

One lovely spring evening in 2023, I pranced out of my house. Babysitter secured, feeling footloose and fancy-free, I was ready for a rousing moms' night out with a friend I hadn't seen in a while. Where was this exciting rendezvous to take place? At the public school down the road. The occasion? A parent-teacher association meeting. Yes, when you have two small children and your husband travels a bunch, even a PTA meeting is something to look forward to. My kids weren't school-age yet. But I wanted to attend because the topic was relevant to their future. The discussion was about whether the local school—where they'll eventually go—might be closed as part of redistricting (that dreaded word!).

A task force had been convened. This meeting was to solicit community input. And so as you might expect, there was no shortage of opinions. Few things get parents as fired up as issues about their children's schooling. Needless to say, the possibility of closing the school had drawn fierce criticism. The school was beloved. Also, most kids would have longer bus rides, and they already had to get up too early. Budding friendships would be bound to wither, as the children would be scattered across different schools. The schools they would be sent to would be overcrowded. And besides, why was *our* school the one being targeted for closure?

What struck me as I entered the building was how big I felt walking down the hall past a water fountain at the height right for a munchkin. The walls were festooned with children's charming drawings, of houses with huge yellow suns hovering over them and stick-figure families smiling and holding hands. Memories of the joy I felt in those years of school washed over me. As I entered the gym, I was warmly greeted by the head of the PTA and an endearing hospitality spread of kid-sized pouches of pretzels and cookies. I spotted my friend and sat down next to her just as a school district representative was about to kick off the meeting.

The district representative was all business, dressed in a skirt suit and sitting ramrod straight in her front-row seat. When the clock turned to 7:00, she stood up and strode to the front of the room, the clack of her heels on the shellacked gym floor breaking through the buzz of parents chatting. She welcomed us and started by emphasizing that the task force was "very open-minded and had not yet decided on a course of action," stressing that they were "truly open to all solutions." That "truly" immediately activated my Spidey sense. She struck me as trying just a tad too hard to convince us that our views were welcomed and would be seriously considered—the way someone who assures you of their honesty causes you to question it (see chapter 2). I wondered if the district administrators had actually already decided to close the school. Others seemed suspicious, too. People shuffled in their seats, and some began whispering to one another.

As someone who spends much of her professional life trying to win over classrooms of students, I felt for the district representative. She was in a tough spot, representing the administration in front of a group of overeducated parents who were cynical, if not outright angry. And as soon as the audience heard her next words, things went completely off the rails.

"Now, before we get to the results of our commissioned enrollment study"—a cornerstone piece of research that would inform the decision of whether to close the school—"I'd like us to do a small group exercise."

Wait, what?

She pointed out that six flip charts were set up around the perimeter of the room and told us to disperse and jot down the school's strengths and weaknesses. Hands shot up, and whispers turned to catcalls.

A man in the second row shouted, "Come on, let's cut to the chase!" A woman sitting in the front row turned around to face the crowd and warned, "It's a trap!"

The district rep was caught off guard, but she was quick to assert her authority. Holding her hands in a "will everyone be quiet" gesture, she said, "I know you might not like it, but we *are* going to do this exercise before continuing." Her tone was unmistakably patronizing.

"What bullshit!" someone exclaimed. Other parents vented, "I don't have time for this crap!" and "We're not third graders." A prominent lawyer in the community piped up. "Where's the transparency and due process?"

My friend and I shot each other "wow, this is getting ugly" glances.

Then something amazing happened.

The school principal intervened. She stood up from her seat in the front row and said, "I know there is a lot of anxiety here. I myself feel very anxious about what will happen." The crowd fell quiet, listening intently now.

"Change is hard, and uncertainty is even harder," she continued. "We all love our school, and we're scared of losing it. We're scared of some of the possibilities that are on the table. We worry about the upheaval of our children's education. I get it. I hear you. I don't know what is going to happen, and that makes me feel out of control and anxious. But I do know that we will come up with a plan. And we will do this with input from all of you. I have confidence in the process."

A wave of calm coursed through the room. Several parents' hands rose, and the principal deftly fielded their questions.

"We're going to have time to share our views, aren't we?" a woman asked.

"Yes, I promise you," the principal assured us.

"Has a decision already been made?" a man challenged.

"No, it hasn't," she said convincingly, followed by: "The task force wants your input, and the exercise we're asking you to do is part of that."

The next thing I knew, parents around the room were getting up and walking over to the flip charts. Soon, everyone was doing the damn exercise. Not even begrudgingly. We did it with gusto.

In the back of my mind, as I wielded a Sharpie and bounced ideas off my newfound collaborators, I kept wondering how the principal had turned things around so fast and so totally. She had opened up and shown her own feelings to a room of ticked-off parents—i.e., she had allowed herself to be vulnerable in public, which, as we know, is a risky move. It's especially risky for leaders because it can threaten perceptions of their competence. "The last time I cried was when I was a baby," Donald Trump famously claimed when he ran for president in 2015.

Herminia Ibarra, a professor of organizational behavior at London Business School, tells the cautionary tale of Cynthia, a leader in a healthcare organization. When Cynthia got a big promotion, the number of people reporting to her increased tenfold. That huge boost of responsibility was daunting to her, and she confided in her staff, "I want to do this job, but it's scary, and I need your help."

"Her candor backfired," Ibarra reports. "She lost credibility."

That's why I was so impressed by the school principal's willingness to be vulnerable in front of a roomful of angry parents, many of whom seemed on the verge of sabotaging the whole meeting. By admitting to feeling "out of control and anxious," she risked being judged as ill-equipped to lead the district through a trying time. But she clearly managed to sidestep the pitfall of appearing incompetent and won over a hostile crowd with her heartfelt speech. I think that she was effective because she shared her own distress and framed it as a shared experience. That approach helped convince people to do the silly flip chart exercise. In contrast, the hard-driving district rep tried to steamroll us into compliance, which alienated her even more.

What would Erving Goffman have said about the school principal's openness, especially in such a public setting? I'm guessing he probably would have disapproved, because he strongly advised putting on a carefully crafted persona in such "front stage" situations. Julianna Pillemer, whose work we encountered in the previous chapter, might consider this exposure of a "sensitive inner state" to be a riskier, level-two form of backstage access.

But the principal had already established herself as a leader. What I haven't yet told you is that the parents *love* this principal, for both the warmth *and* the competence they'd already seen ample evidence of during their children's years at the school. So by the time she displayed vulnerability at this meeting, she already had the public's trust. This gave her more latitude for self-expression—to air her anxieties about the school's fate without fear of undermining her authority. On the contrary, her candor augmented her authority. It made her relatable—something that becomes harder, but more important, as leaders rise through the ranks.

The Warmth/Competence Tightrope

The most inspiring leaders balance warmth and competence. But for women, this balancing act is especially fraught. Research by Susan Fiske and others shows that highly competent women are often perceived as less likable, while warm, approachable women are assumed to be less capable. It's as if we see warmth and competence on a seesaw: When one goes up, the other must come down. For men, that perceived trade-off doesn't exist. We're much more comfortable seeing them as both warm and competent.

That's part of why Cynthia, the healthcare executive we just met, faced backlash. She was new in her role and didn't yet have the trust of her team. So when she admitted to feeling intimidated by the responsibilities ahead, it came off not as relatable but as weak. In trying to humanize herself, she undermined her perceived competence.

The school principal, by contrast, had already built up trust. Parents respected her, and many had known her for years. So when she expressed her anxiety at the PTA meeting, it didn't diminish her authority—it deepened it. Her vulnerability felt grounded in credibility, not insecurity. She was seen as strong enough to be open. And as for the school district rep, well, she tried to assert authority through control, leaning heavily on her rank. But in a room full of emotional, worried parents, what we needed was credible reassurance. Her rigidity came off as cold, even dismissive—and she lost the room.

The same pattern shows up again and again. A friend of mine, Jane, worked under a boss who was highly capable but emotionally withholding. She was directive, unyielding, and rarely invited feedback. Her team was demoralized, and turnover was high.

Then one day, Jane discovered that her boss was the life of the party in her beach town. She cracked jokes, told self-deprecating stories, and had earned a reputation among neighbors as warm and charming. That warmth, however, never made it to the office. In the workplace, she relied on command and control. Eventually, her division was shut down, and the boss was let go.

Stories like these underscore just how hard it is for women leaders to calibrate disclosure. Share too little, and you risk being seen as aloof. Share too much—or at the wrong time—and you risk seeming unqualified. Women are often expected to reveal more, yet are judged more harshly when they do.

So it's no wonder that so many of us, especially when stepping into leadership roles, default to a show of unwavering confidence. We worry that revealing any doubt or weakness could be professionally fatal. But that instinct to overproject confidence can just as easily backfire.

Confidence Flexing

Here's something I see all the time in the classroom: A student takes a bold stance. As discussion unfolds, the evidence clearly contradicts their position. But instead of adjusting, they dig in deeper. Rather than backing down, they double down. In these instances, I often wonder: What does the rest of the class think of this person? Does unyielding confidence inspire trust—or skepticism?

To find out, my colleagues and I—led by Martha Jeong—ran a

series of studies. In one, we analyzed data from an entrepreneurial pitch competition. Think *Shark Tank*: Founders pitch their ideas to a panel of seasoned investors. After the pitch, the judges ask questions and offer feedback. Sometimes these questions reveal serious flaws in the business model—whether around pricing, target audience, or product design. We wanted to know how entrepreneurs respond when challenged like this. Do they acknowledge the critique? Do they let it change their mind? Or do they confidently dismiss it?

The vast majority of entrepreneurs in our study, 76 percent, dug in and defended their ideas. These were the confidence flexers. One entrepreneur, for instance, pitched an online platform to help families hire eldercare providers. He said the target users would be seniors themselves. One investor gently pointed out that it was far more likely their adult children would be the ones seeking care on their parents' behalf. "Why wouldn't you focus on family members?" the investor asked. The entrepreneur replied, "We don't want the focus to be on family members." This wasn't a reason, it was a reflex. And it didn't help him.

What we found was that the entrepreneurs who acknowledged shortcomings were more likely to advance to the next round than those who dismissed the critique. Almost six times more likely. That's a GIANT effect. In short: People who admitted weaknesses in their business plans were rewarded. Those who stubbornly defended bad ideas weren't. In subsequent experiments, we probed into the nature of the impression made by entrepreneurs who express unwavering confidence in their business plans. To put it bluntly, they are seen as confident fools.

That contrast reminded me of the conversation I had with Craig, the executive we met in chapter 9. I asked him what he tells leaders who worry that opening up will make them seem less capable. He grinned and said, "The answer is, almost universally, it's going to be great."

Strength in Weakness

In 2010, a newly minted neuroscientist named Melanie Stefan applied for an early-career grant. When a rejection letter arrived, she was crushed. As it happened, that same day, Brazil's World Cup roster was announced, and to her surprise, the legendary Ronaldinho didn't make the cut. Sports commentators said he was past his prime.

Stefan, a soccer fan, was struck by how when elite athletes fail, their rejections are often public. Professional athletes can't hide their off days or flubbed shots. That visibility, Stefan realized, might actually help others. If even great players struggle, maybe there's value in letting others see your own failures.

So she wrote an article urging academics to create "failure CVs"—lists of the grants, jobs, and papers they didn't get. "Don't dwell on it for hours," she advised, "just keep a running tally." Her point was simple but powerful: Failure is part of the process, and hiding it only reinforces unrealistic standards of perfection.

That idea stuck with me. Could being honest about failures be a kind of leadership?

In a series of studies, my colleagues and I tested how leaders were perceived when they shared their areas of weakness. In one, we asked a Google executive to record two versions of a short introductory video of himself—the kind of thing that might be shown to a group of new hires. In one version, he revealed, truthfully, that he had joined the company only after applying to "thirty-six other roles and consequently receiving thirty-five other rejections." Then we recruited four hundred online participants, all of them full-time professionals, and asked them to imagine that they had just signed on with a new company and were meeting one of the managers. Some of the participants saw the executive's introductory video with his failures included, while others saw it with that part edited out. Those who saw him talking about his failed applications were more motivated to work for him than were the participants who saw the video without the disclosure. In other words, the leader who opened up was more inspiring to followers.

The same was true of leaders who revealed work-related weaknesses. In one study, participants who read about an executive who revealed that he is not good at public speaking, and that it can even make him panicky, were more inspired to work for him relative to when no weakness was revealed. In all of these studies, the results were the same: Opening up about a weakness did not cause participants to see the leaders as less competent. In fact, the would-be employees reported trusting the leaders more.

Prominent business leaders have shown the same pattern. After Zoom was rocked by security concerns early in the pandemic, CEO Eric Yuan didn't deflect. In his first interview, he said, "We moved too fast . . . We had some missteps. We should have done something to enforce passwords." He apologized and outlined how they were fixing the problem. His response was well received. As tech CEO Jim Whitehurst put it, "Being very open about the things I didn't know had the opposite effect than I would have thought. It helped me build credibility." These moments of disclosure work because they communicate something deeper: security. A leader confident enough to admit mistakes doesn't come across as weak—they come across as real. And that makes people want to follow them.

Opening up can also make staff more willing to give tough but honest feedback—something every leader needs, but employees are understandably often reticent to give. "Please tell me how I can improve" rarely puts people at ease. Well, in one study, researchers Constantinos Coutifaris and Adam Grant found that leaders who first shared their own developmental goals (read: weaknesses), such as wanting to be more organized or more confident in public speaking, got significantly more constructive feedback from their employees than leaders who took the standard approach of asking (begging?) people to share their honest feedback. By revealing a weakness, they invited meaningful input—and got exactly the kind of feedback they needed to grow in those very areas.

Disclosure with Discretion

As with any kind of sharing, the point isn't that leaders should reveal everything. Far from it. Disclosing unethical or immoral behavior is tricky and can backfire (best to just not engage in that behavior in the first place!). There are plenty of other things leaders should generally keep to themselves, whether they're managing a team, running for office, or parenting teenagers. So how do we know where the line is?

Disclosure should feel proportionate—not so raw or detailed that others grow uncomfortable or start to question our competence. In one study, my colleagues and I tested the boundary between authentic sharing and TMI. We looked again at leaders disclosing a fear of public speaking. When the description was milder, like "I get nervous before big presentations," the leader came across as authentic but still competent. But when the description was more intense, like "I panic and my mouth goes dry," people started to doubt ability. The general lesson is, be real, but be measured about it, *and* pay attention to your surroundings.

Even Cynthia, the healthcare executive whose disclosure backfired, later reflected on the experience this way: "Being authentic doesn't mean that you can be held up to the light and people can see right through you." She'd learned that effective disclosure has a purpose—and a boundary.

One rule of thumb: Ask yourself how you'd feel hearing the same revelation from your own boss. Would it build trust—or make you uncomfortable?

"The line" depends on context, too. Consider the now-infamous case of former BP CEO Tony Hayward, who, during a press conference about the 2010 Deepwater Horizon disaster, said, "There's no one who wants this thing over more than I do. I'd love my life back." That wasn't an option for the eleven workers killed in the explosion. Under mounting pressure, Hayward stepped down a few months later.

That said, it *can* be powerful—and sometimes absolutely called for—for leaders to reveal difficult emotions they're feeling. This is yet another challenging balancing act.

Expressing Negative Emotions

On my first day teaching at Harvard Business School, I worked hard not to disclose or show the emotion I was feeling: sheer terror. Because, yes, sometimes expressing emotions, especially negative ones, can undermine your influence. It's vital that leaders not let rip with their emotions if they are in a public space with people who don't know them well (or at all). Leadership expert Brené Brown calls that oversharing "floodlighting," and cautions that rather than making you seem relatable, it can cause people to "recoil and shut down." When leaders do express powerful negative emotions like regret, fear, grief, or disappointment, they need to make sure that such sharing is intentional and purposeful.

The need for caution leads some people in leadership positions to routinely suppress such emotions, but that is not always the right thing to do, especially in situations that *warrant* strong emotion. If we don't express it, we risk being perceived as callous and uncaring.

Consider, for example, how Queen Elizabeth was severely criticized for how cold she seemed when Princess Diana died back in 1997. For almost the entire first week after the tragic accident, she was nowhere to be seen. She stayed cloistered at the royal family's Balmoral estate, far from the crowds that had gathered to mourn around Buckingham Palace in the city of London. Newspaper headlines asked, "Where Is Our Queen?" and implored her to "Show Us You Care" and "Speak to Us Ma'am"—show us *something*.

Then, finally, at 6 p.m. on Friday, September 5, five days after the princess's death, Queen Elizabeth did something uncharacteristic. She addressed the nation in a live, and heartfelt, address. Standing in front of a window in the dining room of Buckingham Palace, with throngs of mourners and piles of flowers visible behind her as she spoke, she began by saying:

Since last Sunday's dreadful news we have seen, throughout Britain and around the world, an overwhelming expression of sadness at Diana's death. We have all been trying in our different ways to cope. It is not easy to express a sense of loss, since the initial shock is often succeeded by a mixture of other feelings: disbelief, incomprehension, anger—and concern for those who remain. We have all felt those emotions in these last few days. So what I say to you now, as your Queen and as a grandmother, I say from my heart.

She went on to explain, "This week at Balmoral, we have all been trying to help William and Harry come to terms with the devastating loss that they and the rest of us have suffered."

And although she still didn't really *show* emotion in this address—she didn't tear up, nor was there any waver in her voice—she nonetheless expressed it, through her heartfelt words. And for the characteristically stoic royal family, it was a really big deal. Extremely revelatory. As one onlooker remarked at the time, "I think she said everything that she should have said. I can't think of anything that she left out at all. I think it's completely appropriate."

Accordingly, the public softened; they found comfort knowing that the queen, too, was mourning. Through this address, the queen showed that she wasn't just another aloof rich person; she was, in this moment, a grandmother trying to help her young grandsons process an unimaginable loss. Suddenly the nation could see that she, too, had been grieving. Now her reclusiveness was not only completely understandable but laudable. "She was being a proper granny," her first cousin, Margaret Rhodes, reflected in 2012. "What was the point of bringing the boys down to sit in London with nothing to do but sit there feeling sad about mum?"

Then, the next day, during the funeral procession, the queen did something again uncharacteristic. As the hearse passed, she bowed. "It was not a quick bow, nor a shallow one," wrote one commenter. "The woman accustomed to being bowed by the world now lowered her head and humbly honored the princess. More than anything, it was the bow that broke the fever of anger directed at the queen and her family." Again, opening up in these uncharacteristic ways, mourning publicly, was an incredible act of leadership.

Years later, in 2005, one of Queen Elizabeth's private letters from around the time emerged. It was addressed to Lady Henriette Abel Smith, a close confidante who had written the queen a condolence letter after Princess Diana's death. The letter back to Lady Henriette included a postscript, handwritten by the queen: "I think your letter was one of the first I opened—emotions are still so mixed up but we have all been through a very bad experience!"

Reading that letter, I couldn't help but wonder: What if the queen had said more of this publicly at the time? How much more comfort might the nation have felt? What if she had spoken more openly about the rawness, messiness, and complexity of her emotions?

Still, even those brief glimpses of emotion helped reframe the queen in the public's mind—not as a distant monarch, but as a grieving grandmother. Her vulnerability didn't diminish her stature; it likely strengthened it.

And that's the paradox we often face as leaders. We're taught to contain our emotions, yet it's often their honest expression that moves others most. The key is not just what we feel, but how, when, and why we share it.

Knowing what emotions to disclose at work—and when—is an essential component of emotional intelligence. As Daniel Goleman, who popularized the concept of emotional intelligence, writes, a key aspect of emotional intelligence is "marshaling emotions *in the service of a goal* [my italics]."

In the context of leadership in the workplace, the goals should be related to your team's performance—motivating them, supporting them, inspiring them. And somewhat surprisingly, revealing negative emotions, if done right, can spur much-needed action. Imagine your team has just spent months developing a new service, only to have it flop with the first client because of poor coordination. As the leader, what should you do? It would be a mistake to totally tamp down your frustration and pretend everything is fine. Likewise, it also wouldn't be good to erupt in frustration.

Instead, the better path is what sociologist Arlie Hochschild calls "emotional labor"—managing your feelings intentionally and using them constructively. You might say, "I'm really disappointed in the results, and I know you are, too. We made mistakes. But now we need to figure out where we went wrong and how to fix it." This kind of honest but steady response acknowledges failure *and* invites the team to engage in problem-solving, together.

For Crying Out Loud!

Crying is another long-taboo form of workplace emotion. But it has recently made something of a comeback, not just among employees, but leaders, too. I'm, of course, delighted about this.

For decades, public tears have been seen as disqualifying, especially for men. Presidential hopeful Edmund Muskie reportedly ended his 1972 campaign after reporters said he had cried during a wintry press conference, though he insisted the "tears" had been melting snow (!). House Speaker John Boehner, known for his frequent weeping, was often mocked for it—though his departure in 2015 wasn't directly tied to his tears.

It's long been considered unmanly to cry. (Sadly, IMHO.)

But things began to shift in 2016 when Barack Obama teared up while discussing executive actions on gun control. As he spoke about the children killed at Sandy Hook Elementary, he paused, choked up, and said, "Every time I think of those kids it gets me mad." "This is the most emotion an American president has ever shown on camera," according to Jerald Podair, an associate professor of history at Lawrence University in Wisconsin. "I can almost guarantee that when there is some sort of collage shown of this president's presidency, this one moment will be in there." And the public didn't recoil.

Since then, there have been a lot of male leaders, not just politicians but CEOs and other figures in the business community, crying in public. This was particularly true during the Covid-19 pandemic, when tears became a way for leaders to show both their grief for lives lost and their solidarity with those mourning. Political leaders on both sides of the aisle, including Massachusetts governor Charlie Baker, and Los Angeles mayor Eric Garcetti, were all praised for crying publicly while speaking of the disease's deadly toll. And Marriott International CEO Arne Sorenson (now deceased) was said to have offered a "lesson in leadership" when he had to fight to hold back his tears in a video address to his employees about the economic effects of the pandemic on the company's business—and on the employees themselves, many of whom had been put on temporary leaves of indefinite periods or would soon be facing the loss of their jobs.

In a survey of two thousand C-suite leaders conducted in 2022, the change in attitude was striking. Some 44 percent of the executives surveyed thought crying at work on occasion was perfectly acceptable.

There is, however, one big no-no for leaders when it comes to crying at work: Don't do it in a calculating way, such as trying to soften a decision that causes harm. Braden Wallake, CEO of the sales and marketing firm HyperSocial, learned this lesson the hard way in 2022. After laying off a few employees, he started weeping—and thought it would be a good idea to post a selfie of his tearstained face on LinkedIn. "I just want people to see, that not every CEO out there is cold-hearted," he wrote. The condemnation in the comments section was swift and fierce. One poster accused Wallake of "embarrassing transparent narcissism." Many criticized him for seeming to be more interested in showing off his vulnerability than in helping his employees through a tough time. Clearly, crying, along with other forms of vulnerability and disclosure, won't make you seem like a better leader when you aren't acting like a good leader.

There also continues to be considerable pushback to the idea that it's okay for women leaders to cry in the workplace. When *Huffington Post* reporter Catherine Pearson asked fifteen women leaders in 2014 for their take on crying at work, most were appalled at the very idea. University of Michigan professor emerita Marina Whitman, a former General Motors vice president, said, "If the person you're confronting is male, it provides one more excuse to make him think, 'Isn't that just like a woman?'" Similarly, "Every time I have cried at work, I have regretted it," said MSNBC *Morning Joe* host Mika Brzezinski, who ruefully recalled crying in front of the president of CBS News when she was fired from the network. "When you cry, you give away your power," she concluded. I was reminded of that when I saw her tear up recently on camera, after hearing the eulogy delivered by the parents of a hostage killed in Gaza. No public backlash followed. Instead, many viewers expressed appreciation.

These anecdotes aren't a general hall pass for crying on the job; rather, they reinforce Elizabeth Baily Wolf's advice from the previous chapter: If you do find yourself in tears, it can help to connect them to your emotional investment in the work.

The women leaders Pearson interviewed offered familiar advice: If you're going to cry at work, try to do it in the bathroom. But, of course, that's not always possible (I speak from experience!). So when tears well up, revealing what prompted them can reinforce others' perception that you care deeply about your work and team. Handled thoughtfully, it can push back against the old, false stereotype that emotion ought to undermine competence.

With more women stepping into leadership roles, I'm hopeful that crying won't be seen as incompatible with strength. Crying is, for most of us, the most visibly vulnerable thing we might do in public. This makes me think the old rules about "too much" self-disclosure at work are due for revision. What once felt like too much backstage access might actually be—à la Goldilocks—just right.

If the rules are shifting for how much emotion and vulnerability we accept from leaders, it's worth asking: What happens when leaders open up not just within their teams, but on a bigger, more public, stage? What happens when they share something so personal that it ripples far beyond the walls of an office? When leaders go first—really go first—they can spur societal change. Let's look at what I call a *catalyst confession*.

Ripple Effects of Revealing

On November 7, 1991, a modest, low-ceilinged conference room in the Forum, the Los Angeles Lakers' basketball arena, buzzed with reporters and anticipatory unease. Reporters filled the space, their eyes trained on an empty podium, while security guards scanned the room. One journalist, his hair slicked back and his voice pitched with anxiety, narrated to a live TV camera: "We are still waiting, and obviously, the anticipation is getting worse and worse. We would like to get an answer sooner or later. . . . It is a very tense place."

Suddenly, the room fell quiet as a familiar figure entered: Earvin "Magic" Johnson. The beloved six-foot-nine point guard, nicknamed for his electrifying play and near-constant grin, approached the microphone. As he adjusted it to his towering frame, that grin faltered. His voice, usually smooth and confident, trembled a little: "First of all, let me say . . . good after . . . good after—late afternoon." Then, with a deep breath, he delivered the words that would reverberate around the world.

"Um, because of, uh . . . the HIV virus that I have obtained, um, I will have to retire from the Lakers . . . today."

Magic clarified that he did not have AIDS, but was HIV positive. He expressed his intention to become a spokesperson for HIV awareness and safe sex, saying, "You know, sometimes you're a little naïve about it, and you can think it could never happen to you. I'm going to go on. I'm going to beat it, and I'm going to have fun."

In mere minutes, one of basketball's most iconic stars had retired—not due to injury or age, but because of a virus that, at the time, was wrapped in stigma, fear, and misinformation.

Before Magic's revelation, HIV/AIDS was widely referred to as a "gay disease." Many believed the virus could be transmitted through casual contact, and those diagnosed often faced widespread stigmatization and isolation from their communities. But Magic's announcement shattered misconceptions. He became the most prominent heterosexual man in America to contract HIV, and to talk about it openly.

His courage created what public health experts would later call the "Magic Johnson effect." One study estimated that his disclosure led approximately eight hundred additional men to get tested and discover their HIV status in the months following his announcement, particularly during a nine-to-twelve-month window in 1991–1992.

Magic Johnson's press conference was more than a personal reckoning—it was what I've come to think of as a *catalyst confession*: a deeply personal disclosure by a prominent individual that sparks broader social change. Catalyst confessions have the power to ripple outward, changing public attitudes, inspiring others to speak up, and even, as in Magic's case, saving lives.

Not all of them play out on national television. Like Melanie Stefan, the scientist who shared her "CV of failures" to normalize rejection in academia. She didn't have a global platform or a room full of cameras. But she still hoped her honesty might, in her words, "inspire a colleague to shake off a rejection and start again."

Catalyst confessions come in many forms. Some are quiet and personal, while others are public and playful. And sometimes, they're delivered in evening gowns, with a wink and just the right amount of sass. Earlier in the book, we celebrated Brook Mahealani Lee for striking the perfect Goldilocks balance of self-disclosure—not too much, not too little—when, during the 1997 Miss Universe pageant, she blurted out her delightfully unfiltered answer to the final question: "I would eat everything in the world! You do not understand. I would eat everything twice."

I recently caught up with Brook to reflect on that life-changing moment, nearly thirty years later. She's just as captivating as she was on that stage—warm, vivacious, irreverent, and impossibly charming. There's a generosity in her presence, a kind of sparkling ease, but also a quiet fierceness beneath it.

And cliché as it sounds, she remembers that moment like it was yesterday. She told me what it was like inside the infamous soundproof booth with Miss Trinidad and Tobago—"a vacuum chamber filled with whirring air," she said. "You can't hear a thing, and you're facing the whole auditorium like a mannequin." As she stepped out and heard the question—"If there were no rules for one day . . ."— something clicked. "I swear to God," she said, "it was like the skies opened up. Cherubs around my head. I was like, 'Okay, I'm ready. You guys ready?'"

It was a charming moment of spontaneity in an otherwise highly scripted event. But behind the joke was something more: an act of rebellion. At the time, pageant owner Donald Trump was reportedly considering adding a "weight clause" to contestants' contracts that would allow the organization to fire winners for gaining weight.

And so Brook's answer was both unfiltered *and* political. "It was meant to defend the outgoing title holder and to show up as my authentic self," she told me. "I was kind of done with the fat shaming. And if I were to win, there would be no more of that ridiculous, disrespectful behavior on my watch."

Brook's answer was her way of pushing back—a sly jab at the absurd expectations placed on women to look a certain way, especially in public-facing roles. Her gutsy response didn't transform the beauty industry overnight, but it did spark meaningful conversations about pageant culture and body standards.

Catalyst confessions, like all good leadership, start with courage. They rarely guarantee applause, and they don't always lead to success in the traditional sense. But they do model possibility. They show that vulnerability can shape culture, shift norms, and open doors for others to do the same.

Craig, the mental health advocate we met in chapter 9, knows this well. "The only way we can make things easier for people who are struggling to get help is by having others tell their stories," he told me. "We need people who are willing to speak. And if you're one of those people, let me teach you how to do it safely *and* effectively." His advice for leaders sharing deeply personal stories is pragmatic: Limit the horror to 30 percent, he says—just enough so people can relate and understand. You don't need to dwell on the gory details. The remaining 70 percent, he says, should be focused on the hope. What you've learned, how you're coping, what support looks like. And crucially, he adds, it doesn't have to be a comeback story with a tidy bow. The hope can be in the honesty, the clarity, and the simple act of naming what's hard. That alone can be radical.

And that, in my book, is true leadership—not just inspiring people to follow, but empowering them to reflect, to reveal, and to rise.

Epilogue

n early 2019, my now-husband Colin and I decided to celebrate our recent engagement with an impulsive trip to Rome. I was elated—after a long awkward period of perpetually swiping left (no need to rehash my dating woes; see chapter 8), we had finally found each other! We wanted to spend a weekend sharing our secrets and stories with each other, to learn everything we might have missed in the early stages of our courtship, amid the picturesque anonymity of a foreign country.

Rome proved to be a fitting backdrop. The city's well-preserved ruins attest to how bearing witness to strangers' lives—even across thousands of years—is an essential act of humanity. We rented a cute little love nest close to the Colosseum, romantic balcony included, and went on long walks, interspersed with plenty of picnic breaks to gorge on meats, cheeses, and wine in parks. All the while, we opened up about our pasts even more than we already had, committed contented moments to memory, and imagined the many happy times we would create together.

I know—I know. This is aggressively sappy. A Nicholas Sparks novel would tell me to dial it back. But what can I say? I love love.

And if there's ever a time for unfiltered romance, it's when you're newly engaged in Rome, three Aperol spritzes deep.

My mood dampened only briefly. During an afternoon visit to our tourist-packed neighborhood Colosseum, we happened upon a placard about the practice of *damnatio memoriae*, translated as "damnation of memory"—a punishment in ancient Rome that entailed total erasure of an emperor after their death. Their homes destroyed. History rewritten to exclude them. Statues, portraits, and currency defaced. This legally mandated social death was deemed worse than a physical one. Reading the placard side by side with Colin, I felt a chill deep in my bones as I imagined what it would be like to be banished from memory, willfully forgotten. I thought about our relationship. Not long ago, Colin and I had lost each other and almost erased the other from our lives, all because of our shared fear of expressing our love. Thank goodness he had had the courage to reach out. If he hadn't, the memory of us would have been fixed in time and tinged with "what might have been." Maybe it would have grown too painful to remember. But now we would create a lifetime of memories. I squeezed his hand.

Our desire to escape to Rome to swap secrets clashed jarringly with the ancient punishment we learned about there. Both speak to a fundamental feature of being human: *the desire to be known*, whether in the present or in the memories of the people who knew us. The craving to be authentically known and accepted for who we are is essential to our ability to flourish. It also requires us to be vulnerable—to reveal who we are and how we really feel, even when it seems risky.

Yes, as we've seen, there is such a thing as oversharing. We can reveal too much at the wrong time, with the wrong person, or in the wrong place—and we can suffer hardships when we do. Undeniably, there is real risk to opening up. With the science I have shared in this book, I hope that now we can understand and consider these risks in a more intentional way.

Because with risk come life's greatest rewards. Like finding the love of your life when you were starting to lose hope. Reconnecting with a long-lost friend you thought you'd never speak to again. Finally telling someone the truth about who you are—and being met with acceptance. Or achieving a breakthrough in therapy that shifts something deep inside you. On the topic of sharing our feelings, I can say pretty unequivocally that we should face our fears, or whatever hesitance we feel, and open up more. This is especially true when it comes to expressing our love for others. After all, we don't want to look back on our lives and realize we didn't share as much as we'd wanted.

For eight years, Australian palliative care nurse Bronnie Ware took care of the dying. She grew so close to many of her patients that they opened up about their end-of-life regrets with her. Ware ended up writing a book based on their insights and reflections, entitled *The Top Five Regrets of the Dying*. Their revelations are incredibly poignant—and offer valuable guidance on how to avoid experiencing similar regrets when we are looking back on our own lives. They are:

1. I wish I'd had the courage to live a life true to myself, not the life others expected of me.
2. I wish I hadn't worked so hard.
3. I wish I'd had the courage to express my feelings.
4. I wish I had stayed in touch with my friends.
5. I wish I had let myself be happier.

Notice that, with the exception of number two, all of these top regrets have something in common: They are about things people wish they *had done* rather than things they wish they *hadn't* done. Flash back to chapter 3, where we learned of Thomas Gilovich's work on regret: It's the things we don't do—those "sins of omission"—that end up haunting us the most. This is not just the perspective of an Ivy League professor analyzing data. The same insight emerges from Ware's raw, real-world experiences as a palliative care nurse. Both sources—one steeped in research, the other in intimate, end-of-life conversations—point to the same truth: When looking back, people don't ache over the embarrassing things they did or said nearly as much as the meaningful things they never found the courage to say or do.

Time and again, Ware's patients yearned to have created an authentic, more open, connected, and happier life—changes that could occur only through action, but that often withered away due to inaction or neglect: failing to pursue a passion due to lack of nerve; not telling someone how you feel about them; failing to pick up the phone; not finding ways to overcome doubts and worries. The amount of failure we can rack up in life just from *not doing things* is incredible, when you think about it—and incredibly sad.

Naturally, I want to home in on the third item on the list: regretting not sharing one's feelings. This is the problem of TLI that we've encountered throughout the book. We saw it when John and Grace, our Australian and American lovebirds from chapter 1, spent twenty years apart after lacking the nerve to be fully open about their love for each other (but thankfully took the risk of reconnecting and ended up living happily ever after). We also saw it in my friend Jane's boss who kept her fun-loving sense of humor under wraps at the office and consequently contributed to a demotivated staff. Notably, regrets over things said, those TMI moments that can keep us up at night—embarrassing job interview faux pas, careless words shouted angrily at a spouse—didn't make the list at all.

My collaborator Elinora once told me about a conversation she had with her ninety-two-year-old grandmother during a trip back home to Bulgaria. One thing in particular had earned Ellie a lot of disapproval in recent years—her tattoos. The last time Ellie had been to Bulgaria, she had five or six tattoos. Well, imagine how her stern grandmother reacted when Ellie came back a few years later, this time with more than sixty of them! One night, Ellie decided to confront the issue. "Do you want to know why I have these tattoos?" she asked. Her grandmother listened patiently as Ellie talked about what the tattoos meant to her. One of them marked the spot of her volleyball career-ending injury. One was for her grandfather's favorite tree. A few covered scars she had made herself. In the process, Ellie opened up about things she had been carrying, including a fractured relationship with her father—her grandmother's son. As she opened up, Ellie started crying, and she cried for an hour while her grandmother—normally quick with advice—just listened, holding her in her arms, stroking her hair like when she was a child.

When Ellie finally finished, her grandmother sat in deep thought. And then she acknowledged, out loud, something she had never admitted before—that silence had been passed down through generations, and that maybe it was time to break the cycle.

"You know," she told Ellie, "I wish that I had talked to your father more when he was growing up. We didn't do that, you know? Maybe I should have shared with him more. And when I was a child, it was even worse—no one shared anything. That's just not the way our parents were."

When Ellie told me this story months later, she was still struck by it. She never thought she would be able to have this conversation with her grandmother, much less be met with so much tenderness—and introspection. Sitting in my office, she remembered how safe she had felt curled up in her baba's arms, and she wondered out loud: How many times had she left a moment like this pass her by because she was scared of being vulnerable?

Love, the Most Meaningful Reveal

You never know the power of telling someone you love them. For journalist Laura, being told that may have saved her life.

In 2012, Laura's beloved mother, who had long suffered from depression, died by suicide at the Grand Canyon. The following summer, Laura, then in her forties and the mother of four kids, was so depressed that she herself had been experiencing suicidal thoughts. Impulsively, she bought a same-day one-way ticket to Phoenix. She wanted to be with her mother in the canyon. Crying, she told her children that she "needed to get out of the house for a bit." Her eleven-year-old son, Theo, handed her a note, which she tucked into her purse without reading.

Before Laura got to the airport, she began crying so hard she could barely see. She pulled off into a parking lot, sobbing, unable to return home, but not able to catch her plane, either. Reaching into her purse for a tissue, she found her son's note and read it. Handwritten on an index card, it said, "I know U love me and I love U. Theo."

I can't do this, Laura thought, crying even harder. She could not put her children through what she had been through, she decided. She turned the car around and drove home.

Though Laura is frank about having had suicidal thoughts for many years, knowing how much her family loves her has brought her back from her own ledge time and again. Theo's note has continued to provide comfort. "I carried it in my wallet for years and now keep it on my dresser, a tiny piece of hope and love to see daily," she explains. And unlike her mother, who largely kept her mental health struggles under wraps, Laura has opened up about her most difficult feelings with her therapist and accepted help from her family and friends.

She has also quite consciously taken the risk of being open about her experience in a book and articles about her mom's death. Although Laura understood that telling her story might make potential employers think, "Wow, she's a mess," she ultimately recognized the other side of that equation: "The risk is I don't talk about it and other people don't either." To her, that risk—TLI—was greater.

Sometimes taking the risk of TMI can be an act not just of healing but of great generosity and love, as it has been for Laura.

Consider one of palliative care nurse Bronnie Ware's former patients, Jude, a forty-four-year-old woman dying from motor neuron disease. Jude had been estranged from her mother for years due to choices her mother vehemently opposed. Before Jude's health deteriorated, she decided to write her mother a heartfelt letter, expressing her love and recounting happy memories from her youth.

After receiving the letter, Jude's mother visited her every day until Jude passed. She spoke tenderly to Jude while holding her hand. She apologized for being judgmental of her daughter's life choices and admitted being jealous of her courage in living an authentic life, unbothered by the opinions of others. Tears streamed down Jude's cheeks as her mother spoke.

Later, Jude's words stayed with Ware: "We must learn to express our feelings *now*," she had told her. "Not when it is too late. None of us ever know when it will be too late. Tell people you love them. Tell them you appreciate them. If they can't accept your honesty or [they] react in a different way to how you hoped, it doesn't matter. What matters is you have told them."

Her words stay with me, too. As I was writing this book, something terrifying happened in my own family, an experience that reminded me of the importance of sharing, and especially of sharing our deepest feelings.

It was Christmas Day 2023. As we sat down for the holiday meal, my quite healthy seventy-four-year-old mother suddenly felt very ill and went upstairs to bed. When I went to check on her a little while

later, I found her lying on the bathroom floor, at the base of the porcelain toilet, clearly having vomited. She looked at me foggily, as if
she didn't recognize me. I rushed to her and held her in my arms. As
soon as I did, her eyes rolled back into her head, and she went limp.
I thought she had died.

I yelled to my family to call 911. I ran to the home of a neighbor, who is a nurse. I spotted her through her dining room window, seated with her family. I frantically banged on the window. She
looked up in shock. "It's my mom! She's dying! Help!" I cried desperately. She leapt up and we ran over to my house—to find my mom
sitting up and lucid, with a "what's with all the kerfuffle?" look on
her face.

Soon after, the paramedics arrived. When we described what
had happened, they explained that when a person vomits violently,
as my mom had, it is possible to disrupt the vagus nerve and pass
out. Mom had only fainted, not had a stroke or a brain tumor, as I
had feared.

All's well that ends well, but the experience really shook me. I'm
still haunted by the horrifying image of my mom, eyes rolled back
into her head, unconscious in my arms. It's prompted me to write
down all the things I want her to know. All the ways I love her. All
the ways I'm grateful to her. I'm not counting on her dying anytime
soon. But I'm not taking her for granted, either. I don't want any of
the good stuff left unsaid. Because, at the end of the day, I don't
want to live with the kind of regret that makes memory unbearable.

With my mom, the idea of losing her felt like an unimaginable
shock—sudden, unexpected, over in an instant. With my dad, it's
been something else entirely. A slow, quiet kind of sadness. As an
Alzheimer's patient, he has been losing himself, little by little. He is
slowly forgetting. And in that forgetting, he is also being forgotten—
not by us, but by his own mind. It's one thing to be erased from his-

tory, to be forgotten by others. It's another thing entirely to forget yourself. To me, it is an unthinkably cruel fate, worsened by the fact that, as a lifelong academic, his mind is so tied to his identity. I can hardly grasp how painful that must be for him—especially in his moments of lucidity, when the fog lifts just long enough for him to realize that something is missing. That he is missing. It's the kind of loss that has no clear edges, no single moment you can point to and say *this is when it happened*. It just stretches on—a long, slow disappearing act.

Not long ago, my parents were visiting, and my mom was telling me about a senior living community they'd recently toured. She was explaining how much she liked it—how it seemed like a good fit, how the people seemed wonderful, how nice the grounds were, and how it even had a memory care facility for my dad, when the time comes.

We were sitting around the table, casually chatting, when my dad—who'd been mostly quiet, as he often is these days—suddenly looked up and said, with perfect timing and a sly little smile:

"What did you say it had again?"

Without thinking, my mom and I both leaned in, ready to repeat ourselves—as we so often do these days. That instinct has become automatic: Say it again, slow it down, remind, try not to sound exasperated. (And if I'm being honest, we don't always hide the frustration as well as we mean to.)

But just as we were about to answer, we paused. I looked at him. And realized . . . he was doing a bit. A man with Alzheimer's, feigning forgetfulness. It was one of the clearest, sharpest moments I've had with him in a long time. His mind, even if just for a beat, was completely his. Still witty. Still playful. Still him.

Lately, when I visit, I've been asking him about his life—his childhood, his memories, the things he still holds on to. With Alzheimer's, long-term memory is often the last to go. And so, in these conversations as he shares with me, I get glimpses of the person he was, the person he still is, even as parts of him slip away. Like the time the song "Chantilly Lace" came on the radio and, to my astonishment, he started singing every word. I joined in. For three minutes, we weren't caregiver and patient—we were a father and daughter belting out the Big Bopper in perfect harmony.

Through these visits, I've learned things I never knew: how close he felt to his grandfather—how they would sit in comfortable silence by the fire, just being together. I learned about the miserable seasickness his father endured when they emigrated from Wales to Canada on the *Empress of Australia*. And I learned how, the moment they arrived, his father stood on the dock, took a deep breath of Canadian air, and proclaimed, "This is Canada. We are never going back." A single sentence that marked the start of an entirely new life.

On one of these recent visits, I asked my dad if he had any regrets. Without missing a beat, he said one of his favorite sayings: "Regret is a terrible thing." Then he paused. And paused. And in that pause, I knew.

He didn't have any. Well, not any that really mattered.

We should all be so lucky.

That's my hope for you, too—that you don't look back and wish you had opened up just a little more. That you don't hold back out of fear, only to realize too late that you missed your chance to say the thing that needed to be said. If this book does anything, I hope it helps you make the right choices about what to share, when to share, and with whom. I hope it gives you the courage to take the risks that matter—the ones that lead to deeper connections, richer relationships, and a life you look back on with satisfaction and joy.

Because this is your story. And you are the only one who can make sure it's told.

Acknowledgments

I feel like Elephant and Piggie in *We Are in a Book!*—that brilliantly mcta Mo Willems children's story where they break the fourth wall, realize they're being read, and gleefully exploit it to make the reader say funny words. (Spoiler: Their favorite is "banana.") In their own adorably existential way, they discover what it means to be known—and to know they're being seen. That, in a nutshell, is what this book is about, too: the strange, vulnerable, and sometimes hilarious act of revealing ourselves. Writing it was a very meta process for me—not just because I was constantly thinking about disclosure dilemmas, but because I was constantly living them. Asking myself: Is this too much? Too little? Is this story meaningful or just mortifying? And trying, again and again, to find that sweet spot. Hopefully I mostly got it right. Fortunately, early on, I had my own Elephant and Piggie moment: I realized I wasn't doing this alone.

First and foremost, I thank my (Canadian!) agent, Alison MacKeen. "Agent" doesn't begin to capture it—you've been my book whisperer, editor, sounding board, friend, and, at times, therapist. Thank you for being my sherpa. This book simply wouldn't exist without you. You saw through the academic clutter and worked tirelessly to help shape something clearer, sharper, and more human. Your guidance was steady, your standards high, and your belief in me unwavering. I'm deeply grateful.

Courtney Young, you are a force. The Zen Goddess of publishing. It's been rewarding and humbling working with you—and occasionally being saved from myself by you. I'm sorry for all the em dashes and parentheticals (and also not sorry?).

Celeste Fine, who, along with Alison MacKeen, saved me from writing a boring book on privacy—your badassery both intimidates and inspires me. You could probably run a country, but, fortunately, you chose to help us authors instead.

Thanks also to the dazzling teams at Riverhead, Calligraph, Park Fine & Brower, and beyond. This includes: Nerylsa Dijol, Tom Dussel, Jynne Dilling, Kitanna Hiromasa, Geoff Kloske, Lavina Lee, Corinne Leong, Lauren Peters-Collaer, Laura Rosenblum, Shailyn Tavella, Claire Vaccaro, Michelle Waters, Helen Yentus; Jess Hoare; Kimberly Brower, Abigail Koons; Allyssa Fortunato, Adam Benavides, and Hilary McClellan. I'm also grateful to the visionary Caroline Sutton, for believing in this book from the get-go and for invaluable feedback.

Thank you to those who shared your stories with me. Your openness—past and present—helped bring these pages to life. Thank you to the talented editors who helped me shape my ideas and tune my writing for nonacademic readers. Emily Loose, for giving me early confidence and teaching me how to structure a chapter. The ruthlessly brilliant Beth Rashbaum, who lovingly bullied me into bringing the book to life through stories instead of statistics. Aaron Shulman, who made everything sharper, stronger, better. Katie Shonk, a straight shooter and incredible detail-wrangler who helped me with everything from fact-checking to inertia-busting. Thank you also to Pete Garceau, for the gorgeous cover art, and to Dan Meretzky, for bringing the visuals in the book to life.

My beloved Pittsburgh sisters—Cindy Cryder, Tamar Krishnamurti, Kim Murtaugh—you always show up. Here, you read the manuscript cover to cover and gave feedback that was unflinchingly kind and piercingly smart—just like you. You definitely made the book better. Someday we'll realize our dream of communal family living (now that that dream is officially on the record!). And to Paul Litvak (can I call you an honorary Pittsburgh sister?), endlessly insightful and early reader par excellence. All that being said, any lingering errors are, of course, my own.

Bigger picture, I'm lucky to have a job where I'm surrounded by minds I admire. I'm fortunate to call HBS my professional home, a place of deep learning, steady support, and creative freedom. Apologies, this is going to be a bit of a long list, but I don't think I could have developed the ideas in this book without the influences of the following individuals. (Instructor hat, just for a sec: Chapter 5 reminded me how easy it is to undershare appreciation; this list is me trying not to.)

First, I'm deeply indebted to my besties Alison Wood Brooks and Mike Norton. You inspire me and embody the "our jobs are absurd and amazing" ethos like no one else. Mike, thank you for never making me feel stupid—not even when I showed up at your office in tears, unannounced, overwhelmed by some now-forgotten existential crisis. And not even when I use my fingers to count to ten. Alison, thank you for helping me through umpteen life transitions. I've learned so much from both of you—about research, about friendship, and about the joy of not taking yourself too seriously.

To my research assistants: Elinora Pentcheva, research assistant turned friend and true thought partner. Thank you for your openness, for your stats wizardry, and for helping me stumble through this process with some semblance of grace. You've taught *me* both savoir faire and savoir être. Anne Marie Green: perceptive, incisive, and unfailingly positive. And to the many others whose behind-the-scenes work laid the foundation and beyond: Rebecca Altholz, Marina Burke, Louisa Dillon, Claire Dirks, Jeff Lin, Ivor Mills, Dora Nathan, Shannon Sciarappa, Trevor Spelman, Simon Wesenberg, and Songyang Zhang.

Todd Rogers and Ryan Buell—talking ideas with you is magic. You also introduced me to the world of Mo Willems via baby gifts for our firstborn, setting the tone early for the playful, thoughtful parenting (and literature!) we aspire to.

To my cherished collaborators and doctoral students I had the joy of working with on the research featured in this book. You make me think *and* laugh. Alessandro Acquisti, Gabe Adams, Kate Barasz, Hayley Blunden, Kristina Brant, Aaron Brough, Hanne Collins, Katy DeCelles (I promise not to send you lewd gifts anymore), Grant Donnelly, Oliver Emrich, Ximena Garcia Rada, Eben Harrell, Reto Hofstetter, Holly Howe, Martha Jeong, Li Jiang, Tami Kim, Maryam Kouchaki, Gabriela Kunath, Heidi Liu, Bhavya Mohan, Jimin Nam, David Norton, Ed O'Brien, Mario Small, Ting Zhang, and Joachim Vosgerau (cheers to more la dolce vita moments on Lake Como).

And I can't *not* acknowledge the one and only George Loewenstein, who shaped my taste in ideas—perhaps the single greatest gift a doctoral adviser can give. I'll also never forget how, for our first paper together, you opened a draft I'd rewritten many times, to no improvement, and made me justify every word of the first sentence. We didn't get past the first paragraph but it forever changed how I approach writing. Also please consider this paragraph my formal apology for teasing you in this book.

Thank you to the many mentors, colleagues, students, and friends who have offered me invaluable support and advice of all kinds for this book: Teresa Amabile, Kassandra Brabaw, Zoe Chance, Rosalind Chow, Dolly Chugh, Helena Dea Bala, Mads Faurholt, Alison Fragale, Marisa Franco, Amy Gallo, Allison Gross, Hal Hershfield, Eva Jannotta, Brian Kenny, Leidi Klotz, Ethan Kross, Katy Milkman, Benoît Monin, N.E.R.D. Lab, George Newman, Dave Nussbaum, Team Onagadori, David Pizarro, Kathleen Vohs, Lindsay Ratowsky, Kim Scott, Mitch Weiss, Elliot Williams, and Annie Wilson. Thank you for your wisdom, encouragement, and generosity.

Now, the smoochy personal life stuff! First, a big shout-out to our loving nanny, Kaylee Klenk, and our small army of babysitters—without whom no work would ever get done (and I would lose my mind).

I owe so much to my family. And not just for giving me all the grist on decision-making quirks (though thanks for that, too). To my mom, I'm moved by your openness in letting me share our story in chapter 3. To my brothers, Will and Greg—proof that you can be siblings *without* rivalry. I can't imagine going through life without you. To my dad, fellow academic and joker extraordinaire. Without you, I would never have known this bizarre slice of paradise—and occasional purgatory (peer review says hi)—that is academia. You've written more books than I can count on my hands. I now have a new appreciation for what that takes. I don't know what you'd think of this one, but it's for you. You used to tease me for studying arcane little things—and, in fairness, also steered me to be more accessible. So thank you for modeling both rigor and ridiculousness. And as you and Mom used to say, "If we had bet money on which of our kids would become a Harvard professor, we'd have lost bigtime"—well, you may have lost the bet, but we won the family lottery.

And most important to me: Colin. I'm ever grateful we swiped right—twice. Thank you for going on this book journey with me—for your willingness to be a little more open, and for letting me share our story. To my darling boys, Oliver and Tyler: You are my raison d'être. Ain't nothing more grounding than wiping your tiny little bums.

Notes

1: The Surprising Power of Opening Up

2 **Take John and Grace:** Francesca Street, "They Fell in Love in the 1980s but Married Other People. 23 Years Later They Reconnected," CNN, November 17, 2022, cnn.com/travel/article/chance-encounters -reunited-couple.

4 **"It literally made us cry":** Street, "They Fell in Love."

6 **But sometimes they're trickier:** Ipsita Kaul, "Did You Know That Lara Dutta Had the Highest Ever Individual Score at Miss Universe 2000?," *Elle*, April 16, 2024, elle.in/lara-dutta-miss-universe-2000.

6 **"Now the tension really":** "1997 Miss Universe Top 3 Questions," Miss Universe Facebook page, July 5, 2024, facebook.com/watch/?v=2480116 625500689.

8 **So I showed this clip:** Leslie John and Elinora Pentcheva, unpublished data, 2024.

8 **The findings of my:** This effect holds controlling for respondent nationality.

9 **The results were extremely consistent:** Leslie K. John, Alessandro Acquisti, and George Loewenstein, "Strangers on a Plane: Context-Dependent Willingness to Divulge Sensitive Information," *Journal of Consumer Research* 37, no. 5 (February 2011): 858–73.

12 **As my and others' research:** Emmi Ignatius and Marja Kokkonen, "Factors Contributing to Verbal Self-Disclosure," *Nordic Psychology* 59,

no. 4 (2007): 362–91; Valerian J. Derlega and Alan L. Chaikin, "Privacy and Self-Disclosure in Social Relationships," *Journal of Social Issues* 33, no. 3 (1977): 102–15; Tami Kim, Kate Barasz, and Leslie K. John, "Consumer Disclosure," *Consumer Psychology Review* 4, no. 11 (2020): 59–69.

12 **Sharing more freely can increase:** James W. Pennebaker and Cindy K. Chung, "Expressive Writing: Connections to Physical and Mental Health," in *The Oxford Handbook of Health Psychology*, ed. Howard S. Friedman (Oxford University Press, 2011; online edition, Oxford Academic, September 18, 2012).

2: Why We Stay Silent

15 **In his 1970s campus novel:** David Lodge, *Changing Places* (Penguin, 1979).

16 **He "slammed his fist":** Lodge, *Changing Places*, 136.

16 **He and his colleagues:** Michael L. Slepian, Jinseok S. Chun, and Malia F. Mason, "The Experience of Secrecy," *Journal of Personality and Social Psychology* 113, no. 1 (2017): 1–33.

16 **They distilled those responses:** The thirty-eight were arrived at through iteration. A second, independent sample of one thousand participants validated the coding system (agreement = 84 percent).

19 **"bright but strange":** Sherri Cavan, "When Erving Goffman Was a Boy: The Formative Years of a Sociological Giant," *Symbolic Interaction* 37, no. 1 (February 2014): 41–70.

19 **His full name is:** Erving Goffman, *The Presentation of Self in Everyday Life* (Doubleday Anchor, 1959).

19 **A mother sent her son:** Erving Goffman, "Communication Conduct in an Island Community" (PhD diss., University of Chicago, 1953), library .oapen.org/bitstream/handle/20.500.12657/60113/goffman-1953 -communication-conduct.pdf?sequence=12&isAllowed=y.

19 **Canada itself had turned away:** "Canada Apologises for Turning Away Jewish Refugee Ship in 1939," BBC, November 7, 2018, bbc.com/news /world-us-canada-46105488.

19 **By all accounts, he lived:** Dmitri N. Shalin, "Interfacing Biography,

Theory and History: The Case of Erving Goffman," *Symbolic Interaction* 37, no. 1 (February 2014): 2–40.

19 **At his famously grueling:** Yves Winkin, "The Cradle: Introduction to the Mediastudies.press Edition," in *Communication Conduct in an Island Community*, mediastudies.press, December 4, 2022, doi.org/10.32376 /3f8575cb.21a77b51.

20 **He joined a long tradition:** Khawlah bint Yahya, "Mind Your Tongue: How to Tame a Tiger and Make It Work for Good," Understand Al Quran Academy, understandquran.com/mind-your-tongue-how-to-tame -a-tiger-and-make-it-work-for-good; "Silence Is Golden . . . ," *Hinduism Today*, June 1, 1997, hinduismtoday.com/magazine/june-1997/1997 -06-silence-is-golden.

20 **"Real life consists of bluffing":** Keith Romer, "How AI Conquered Poker," *New York Times*, January 18, 2022, nytimes.com/2022/01/18 /magazine/ai-technology-poker.html#:~:text=%E2%80%9CReal %20life%20consists%20of%20bluffing,Neumann%20thought%2C %20was%20like%20poker.

20 **Game theory helps us navigate:** John von Neumann and Oskar Morgenstern, *Theory of Games and Economic Behavior* (Princeton University Press, 1944).

21 **Consider a shocking:** Andrea Gurmankin Levy, Aaron M. Scherer, Brian J. Zikmund-Fisher, Knoll Larkin, Geoffrey D. Barnes, and Angela Fagerlin, "Prevalence of and Factors Associated with Patient Nondisclosure of Medically Relevant Information to Clinicians," *JAMA Network Open* 1, no. 7 (2018): e185293.

21 **American Academy of Family Physicians:** Dennis Thompson, "Most Americans Lie to Their Doctors," *Medical Xpress*, December 4, 2018, medicalxpress.com/news/2018-12-americans-doctors.html.

21 **Another surgeon recounted:** Maria Zaldivar, "Doctors Share the Most Dangerous Lies Their Patients Have Ever Told Them," *Comic Sands*, October 30, 2019, accessed July 21, 2025, comicsands.com/patients-lie -to-doctors.

23 **In a classic article:** Garrett Hardin, "The Tragedy of the Commons," *Science* 162, no. 3859 (1968): 1243–48.

25 **In Ostrom's seminal book:** Elinor Ostrom, *Governing the Commons:*

The Evolution of Institutions for Collective Action (Cambridge University Press, 1990).

26 **One especially hopeful:** Rebecca Koomen and Esther Herrmann, "An Investigation of Children's Strategies for Overcoming the Tragedy of the Commons," *Nature Human Behaviour* 2, no. 7 (2018): 348–55.

26 **In Ostrom's research, as well as:** Ostrom, *Governing the Commons*; Koomen and Herrmann, "An Investigation of Children's Strategies."

27 **My collaborators, Kate Barasz, Mike Norton, and I:** Leslie K. John, Kate Barasz, and Michael I. Norton, "Hiding Personal Information Reveals the Worst," *Proceedings of the National Academy of Sciences* 113, no. 4 (January 11, 2016): 954–59.

29 **Assurances alone don't:** Christophe Boone, Benoit De Brabander, and Arjen van Witteloostuijn, "The Impact of Personality on Behavior in Five Prisoner's Dilemma Games," *Journal of Economic Psychology* 20, no. 3 (1999): 343–77; Eleanor Singer, Hans-Jürgen Hippler, and Norbert Schwarz, "Confidentiality Assurances in Surveys: Reassurance or Threat?," *International Journal of Public Opinion Research* 4, no. 3 (Autumn 1992): 256–68.

29 **In one study, we asked:** Leslie John, Alessandro Acquisti, and George Loewenstein, unpublished data.

30 **when it comes to trust:** Roger C. Mayer, James H. Davis, and F. David Schoorman, "An Integrative Model of Organizational Trust," *Academy of Management Review* 20, no. 3 (1995): 709–34.

30 **That may sound obvious:** Don A. Moore and Paul J. Healy, "The Trouble with Overconfidence," *Psychological Review* 115, no. 2 (2008): 502–17.

30 **By contrast, actual vulnerability:** Robert H. Frank, *Passions Within Reason: The Strategic Role of the Emotions* (W. W. Norton, 1988).

30 **As Shakespeare put it:** William Shakespeare, *Hamlet*, act 3, scene 2.

30 **That's because, in his view:** Tim Harford, "A Beautiful Theory," *Forbes*, December 14, 2006, forbes.com/2006/12/10/business-game-theory-tech -cx_th_games06_1212harford.html.

32 **Often, though, we perceive:** Max H. Bazerman and Margaret A. Neale, "Heuristics in Negotiation: Limitations to Dispute Resolution Effectiveness," in *Negotiations in Organizations*, ed. Max H. Bazerman and Roy Lewicki (Sage, 1983), 51–67.

3: Understanding Undersharing

35 **For decades, Marilyn Mach:** Julie Baumgold, "In the Kingdom of the Brain," *New York*, February 9, 1989.

36 **In 1989, the cover of *New York*:** Baumgold, "In the Kingdom."

36 **In short, the reader asked:** Marilyn vos Savant, "Game Show Problem," *Parade*, 1990/1991, web.archive.org/web/20100310140547/http://www.marilynvossavant.com/articles/gameshow.html.

37 **"You blew it":** Vos Savant, "Game Show Problem." Found here: behavioralscientist.org/steven-pinker-rationality-why-you-should-always-switch-the-monty-hall-problem-finally-explained.

38 ***Zonks*, which are dud prizes:** After tapings, people who were Zonked were offered a "consolation prize" of about one hundred dollars cash instead of taking home the actual Zonk. "Zonk," Fandom, gameshows.fandom.com/wiki/Zonk, accessed April 17, 2025.

39 **"It's been an intense":** John Tierney, "Behind Monty Hall's Doors: Puzzle, Debate and Answer?," *New York Times*, July 21, 1991, nytimes.com/1991/07/21/us/behind-monty-hall-s-doors-puzzle-debate-and-answer.html.

40 **even when people are told the math:** Stefan Krauss and X. T. Wang, "The Psychology of the Monty Hall Problem: Discovering Psychological Mechanisms for Solving a Tenacious Brain Teaser," *Journal of Experimental Psychology: General* 132, no. 1 (2003): 3–22; Bruce D. Burns and Mareike Wieth, "The Collider Principle in Causal Reasoning: Why the Monty Hall Dilemma Is So Hard," *Journal of Experimental Psychology: General* 133, no. 3 (2004): 434–49.

41 **At the crux of this asymmetry:** Mark Spranca, Elisa Minsk, and Jonathan Baron, "Omission and Commission in Judgment and Choice," *Journal of Experimental Social Psychology* 27, no. 1 (1991): 76–105.

46 **Looking back, she could see:** Raymond S. Nickerson, "Confirmation Bias: A Ubiquitous Phenomenon in Many Guises," *Review of General Psychology* 2, no. 2 (1998): 175–220.

50 **Psychologists Thomas Gilovich:** Thomas Gilovich and Victoria Husted

Medvec, "The Temporal Pattern to the Experience of Regret," *Journal of Personality and Social Psychology* 67, no. 3 (1994): 357–65.

4: Are You a Revealer or a Concealer?

54 **"Boys' emotional expressions decreased":** Tara M. Chaplin, "Gender and Emotion Expression: A Developmental Contextual Perspective," *Emotion Review* 7, no. 1 (January 2015): 14–21.

54 **The study that shook me:** Ross Buck, "Nonverbal Communication of Affect in Preschool Children: Relationships with Personality and Skin Conductance," *Journal of Personality and Social Psychology* 35, no. 4 (1977): 225–36. Although Buck's study involved a small sample—just twenty-four preschoolers—his findings replicated earlier work and have since been echoed by a large body of research on emotion suppression, such as: J. J. Gross and R. W. Levenson, "Hiding Feelings: The Acute Effects of Inhibiting Negative and Positive Emotion," *Journal of Abnormal Psychology* 106, no. 1 (1997): 95–103; Thomas L. Webb, Eleanor Miles, and Paschal Sheeran, "Dealing with Feeling: A Meta-Analysis of the Effectiveness of Emotion Regulation Strategies," *Psychological Bulletin* 138, no. 4 (2012): 775–808.

55 **Rather than going down:** Leslie R. Brody, "The Socialization of Gender Differences in Emotional Expression: Display Rules, Infant Temperament, and Differentiation," in *Gender and Emotion: Social Psychological Perspectives*, ed. Agneta H. Fischer (Cambridge University Press, 2000), 24–47.

58 **Goffman, by many accounts:** Dmitri N. Shalin, "Interfacing Biography, Theory and History: The Case of Erving Goffman," *Symbolic Interaction* 37, no. 1 (February 2014): 2–40.

58 **He "presented himself":** Gary T. Marx, "Role Models and Role Distance: A Remembrance of Erving Goffman," *Theory and Society* 13, no. 5 (September 1984): 649–62.

58 **an "autobiographical reticence":** Michael Delaney, "Goffman at Penn: Star Presence, Teacher-Mentor, Profaning Jester," *Symbolic Interaction* 37, no. 1 (2014): 87–107.

58 **As a young man:** Marty Jourard, *A Curious Mind: The Life and Legacy of Sidney Jourard* (pub. by author, 2020), chapter 3.

59 **Jourard died young:** Earl C. Brown, "Sidney Marshall Jourard (1926–1974)," unpublished eulogy, December 23, 1974, sidneyjourard.com /sideulogy.htm.

59 **He gave public lectures barefoot:** Jourard, *A Curious Mind*, chapter 3.

59 **Jourard's intellectual curiosity:** Body image: S. M. Jourard and P. F. Secord, "Body Size and Body-Cathexis," *Journal of Consulting Psychology* 18, no. 3 (1954): 184; psychedelic drugs, see chapter 6; patient care, S. M. Jourard, "How Well Do You Know Your Patients?," *American Journal of Nursing* 59, no. 11 (1959): 1568–71.

59 **A young Sid:** Marty Jourard, *The Life and Legacy of Sidney Jourard*, "Opening Pages: A Curious Mind," accessed June 5, 2025, sidneyjou rard.com/OpeningPagesCuriousMind.pdf.

59 **"I became fascinated with self-disclosure":** Sidney M. Jourard, *The Transparent Self* (Van Nostrand Reinhold, 1971), 5.

61 **Now that we've explored:** Elinora Pentcheva and Leslie John, "The Agreeable Revealer: Personality Correlates of Self-Disclosure" (work-ing paper). We focus on the Big Five because it's the most widely used, empirically supported, and cross-culturally validated framework for studying personality in psychology.

61 **Extraversion is also the most:** Oliver P. John and Sanjay Srivastava, "The Big Five Trait Taxonomy: History, Measurement, and Theoreti-cal Perspectives," in *Handbook of Personality: Theory and Research*, 2nd ed., ed. Lawrence A. Pervin and Oliver P. John (Guilford Press, 1999), 102–38.

63 **"A person will permit himself":** Sidney Jourard, *Self-Disclosure: An Experimental Analysis of the Transparent Self* (Wiley-Interscience, 1971), 5.

64 **Agreeableness shapes how we weigh:** Agreeableness is by far the stron-gest and most consistent predictor of disclosure optimism, typically ex-plaining about 10 percent of the variance. Neuroticism also predicts it negatively, but much more weakly (about 3 percent). Extraversion has a small, inconsistent effect, while openness and conscientiousness show

minimal and unreliable associations. Pentcheva and John, "The Agreeable Revealer."

65 **Attachment styles refer to:** John Bowlby, *Attachment and Loss*, vol. 1, *Attachment* (Basic Books, 1969); Kelly A. Brennan, Catherine L. Clark, and Phillip R. Shaver, "Self-Report Measurement of Adult Attachment: An Integrative Overview," in *Attachment Theory and Close Relationships*, ed. Jeffry A. Simpson and W. Steven Rholes (Guilford Press, 1998).

65 **There are three basic attachment:** Researchers increasingly conceptualize adult attachment not as fixed types but as two dimensions: anxiety (fear of rejection) and avoidance (discomfort with closeness). "Secure" generally means low on both, though some argue it also involves a distinct capacity for trust and intimacy. Brennan et al., "Self-Report Measurement of Adult Attachment," 46–76; Mario Mikulincer and Phillip R. Shaver, *Attachment in Adulthood: Structure, Dynamics, and Change*, 2nd ed. (Guilford Press, 2017).

65 **To home in on your:** Cindy Hazan and Phillip R. Shaver, "Romantic Love Conceptualized as an Attachment Process," *Journal of Personality and Social Psychology* 52, no. 3 (1987): 511–24.

66 **As with the Big Five:** Pentcheva and John, "The Agreeable Revealer."

67 **After all, Stanley Milgram's:** Stanley Milgram, *Obedience to Authority: An Experimental View* (Harper & Row, 1974).

67 **Understand the people first:** Robert Sommer, "Sociofugal Space," *American Journal of Sociology* 72, no. 6 (May 1967): 654–60; Robert Sommer, *Personal Space: The Behavioral Basis of Design* (Prentice-Hall, 1969).

69 **By comparison, Africa:** Image adapted from Schmitt et al., "The Geographic Distribution," 173-A.

71 **Personal sharing inherently:** William B. Gudykunst and Stella Ting-Toomey, *Culture and Interpersonal Communication* (Sage, 1988).

71 **Psychologists call this *disclosure flexibility*:** Gordon J. Chelune, "Disclosure Flexibility and Social-Situational Perceptions," *Journal of Consulting and Clinical Psychology* 45, no. 6 (1977): 1139–43.

73 **One friend later wrote:** Brown, "Sidney Marshall Jourard."

74 **After his sudden death:** Sadie Brown, "The Jourard Family History," jourard.com/jourardhistory.htm.

5: The Why of Disclosure Decisions

75 **Paul was the baby:** The information about Paul and his family in this chapter is based on my interview with him on September 25, 2024, as well as an article he wrote: "A Dead Brother, a Mystery, a Family Reunion—and the Question He Longed to Ask," *Sydney Morning Herald*, March 11, 2022, smh.com.au/lifestyle/life-and-relationships/over -my-dead-brother-an-uncomfortable-family-truth-is-revealed-2021 1122-p59b2x.html.

78 **But she paltered:** Todd Rogers, Richard Zeckhauser, Francesca Gino, Michael I. Norton, and Maurice E. Schweitzer, "Artful Paltering: The Risks and Rewards of Using Truthful Statements to Mislead Others," *Journal of Personality and Social Psychology* 112, no. 3 (December 2016): 456–73.

81 **In 1772, Benjamin Franklin:** Benjamin Franklin, "From Benjamin Franklin to Joseph Priestley, 19 September 1772," Founders Online, National Archives, founders.archives.gov/documents/Franklin/01-19-02 -0200.

82 **"System 1 thinking":** Keith E. Stanovich and Richard F. West, "Individual Differences in Reasoning: Implications for the Rationality Debate," *Behavioral and Brain Sciences* 23, no. 5 (2): 645–65.

82 **This shift matters:** Katherine L. Milkman, Dolly Chugh, and Max H. Bazerman, "How Can Decision Making Be Improved?," *Perspectives on Psychological Science* 4, no. 4 (2009): 379–83; Valerie F. Reyna and Frank Farley, "Risk and Rationality in Adolescent Decision Making: Implications for Theory, Practice, and Public Policy," *Psychological Science in the Public Interest* 7, no. 1 (2006): 1–44.

82 **In the 1970s, Irving Janis:** Irving L. Janis and Leon Mann, *Decision Making: A Psychological Analysis of Conflict, Choice, and Commitment* (Free Press, 1977).

83 **The truth is, these tools:** Susan E. Collins, Megan Kirouac, Melissa A. Lewis, Katie Witkiewitz, and Kate B. Carey, "Randomized Controlled Trial of Web-Based Decisional Balance Feedback and Personalized

Normative Feedback for College Drinkers," *Journal of Studies on Alcohol and Drugs* 75, no. 6 (November 2014): 982–92.

83 **"The aim is not to challenge":** William R. Miller, "Motivational Interviewing with Problem Drinkers," *Behavioural Psychotherapy* 11, no. 2 (1983): 152.

85 **In a landmark study, Daniel Wegner:** Daniel M. Wegner, David J. Schneider, Samuel R. Carter III, and Teri L. White, "Paradoxical Effects of Thought Suppression," *Journal of Personality and Social Psychology* 53, no. 1 (1987): 5–13.

85 **In a follow-up study:** Julie D. Lane and Daniel M. Wegner, "The Cognitive Consequences of Secrecy," *Journal of Personality and Social Psychology* 69, no. 2 (1995): 237.

85 **Meanwhile, other research speaks:** Anita E. Kelly, "Revealing Personal Secrets," *Current Directions in Psychological Science* 8, no. 4 (August 1999): 105–9; Michael L. Slepian and Katharine H. Greenaway, "The Benefits and Burdens of Keeping Others' Secrets," *Journal of Experimental Social Psychology: General* 78 (2018): 220–32.

85 **As James Joyce put it:** James Joyce, *Ulysses* (Shakespeare and Company, 1922), 28.

86 **Psychologists Daniel Gilbert:** Daniel T. Gilbert and Jane E. J. Ebert, "Decisions and Revisions: The Affective Forecasting of Changeable Outcomes," *Journal of Personality and Social Psychology* 82, no. 5 (2002): 503–14.

86 **In one striking study:** Dylan M. Smith, George Loewenstein, Aleksandra Jankovic, and Peter A. Ubel, "Happily Hopeless: Adaptation to a Permanent, but Not to a Temporary, Disability," *Health Psychology* 28, no. 6 (November 2009): 787–91.

96 **In a phenomenon called:** Leslie K. John, Hayley Blunden, and Heidi Liu, "Shooting the Messenger," *Journal of Experimental Psychology: General* 148, no. 4 (2019): 644–66; Bertram Gawronski and Eva Walther, "The TAR Effect: When the Ones Who Dislike Become the Ones Who Are Disliked," *Personality and Social Psychology Bulletin* 34, no. 9 (2008): 1276–89; John J. Skowronski, Donald E. Carlston, Lynda Mae, and Matthew T. Crawford, "Spontaneous Trait Transference: Communicators Take on the Qualities They Describe in Others," *Journal of Personality and Social Psychology* 74, no. 4 (1998): 837–48.

96 **Research finds that people:** Erica J. Boothby and Vanessa K. Bohns, "Why a Simple Act of Kindness Is Not as Simple as It Seems: Underestimating the Positive Impact of Our Compliments on Others," *Personality and Social Psychology Bulletin* 47, no. 5 (2021): 826–40; Xuan Zhao and Nicholas Epley, "Insufficiently Complimentary?: Underestimating the Positive Impact of Compliments Creates a Barrier to Expressing Them," *Journal of Personality and Social Psychology* 121, no. 2 (2021): 239–56.

99 **Research on procedural justice:** Tom R. Tyler, *Why People Obey the Law* (Yale University Press, 1990).

6: The Healing Power of Revealing

101 **On the morning of:** Royal Canadian Air Force Casualties Officer to Mr. M. A. Duquette, telegram, May 10, 1943, originator number M9867, sent by M. Cameron A/S/O R.O.4, Duquette Family Papers, private collection.

102 **The Duquettes were a well-known:** "Five Duquette Brothers in War," *North Bay Daily Nugget*, November 25, 1942.

103 **And whatever hope:** J. Buchanan to Mr. and Mrs. M. A. Duquette, letter, May 11, 1943, 272 Squadron, Middle East, Duquette Family Papers, private collection.

103 **His Beaufighter had run:** "Report on Flying Accident or Forced Landing Not Attributable to Enemy Action," RAF Form 765C, dated May 11, 1943, Duquette Family Papers, private collection.

103 **A search team spotted:** "229 Squadron Operational Record Book, Form 540," May 8, 1943, AIR27/1419, Duquette Family Papers, private collection; William G. Bromhead, "Logbook Entry, May 8, 1943," 229 Squadron, Duquette Family Papers, private collection.

103 **Fred and Albert were:** W. R. Gunn, "Letter to Mr. M. A. Duquette," Ottawa, November 11, 1943, Duquette Family Papers, private collection.

103 **An inventory of Fred's:** Flight Lieutenant, President, Standing Committee of Adjustment, Army Headquarters, Malta, *Inventory of the Personal Effects of J.178245 P.O. F. A. Duquette*, May 31, 1945, Duquette Family Papers, private collection.

104 **Maybe they shared stories:** Mrs. Johnson to Bert Duquette, letter, November 2, 1943, Duquette Family Papers, private collection.

104 **In Pennebaker's studies:** James W. Pennebaker and Sandra K. Beall, "Confronting a Traumatic Event: Toward an Understanding of Inhibition and Disease," *Journal of Abnormal Psychology* 95, no. 3 (August 1986): 274–81.

104 **Across dozens of studies:** A meta-analysis found that expressive writing has a small but clinically meaningful effect on well-being—roughly twice the size of aspirin's effect in preventing heart attacks. As the author notes, given that disclosure is free, low-risk, and often perceived as helpful, even modest benefits are worth noting. Joanne Frattaroli, "Experimental Disclosure and Its Moderators: A Meta-Analysis," *Psychological Bulletin* 132, no. 6 (November 2006): 823–65. Another meta-analysis found similar results: Pasquale Frisina, Joan C. Borod, and Stephen J. Lepore, "A Meta-Analysis of the Effects of Written Emotional Disclosure on the Health Outcomes of Clinical Populations," *Journal of Nervous and Mental Disease* 192 (2004): 629–34.

104 **HIV-positive patients randomized:** Keith J. Petrie, Iris Fontanilla, Mark G. Thomas, Roger J. Booth, and James W. Pennebaker, "Effect of Written Emotional Expression on Immune Function in Patients with Human Immunodeficiency Virus Infection: A Randomized Trial," *Psychosomatic Medicine* 66, no. 2 (March–April 2004): 272–75.

104 **University students who wrote:** Pennebaker and Beall, "Confronting a Traumatic Event."

104 **Recently unemployed adults:** Stefanie P. Spera, Eric D. Buhrfeind, and James W. Pennebaker, "Expressive Writing and Coping with Job Loss," *Academy of Management Journal* 37, no. 3 (1994): 722–33.

105 **Men tend to get:** Joshua M. Smyth, "Written Emotional Expression: Effect Sizes, Outcome Types, and Moderating Variables," *Journal of Consulting and Clinical Psychology* 66, no. 1 (1998): 174–84.

105 **"Anything that's human":** Fred Rogers, *You Are Special: Words of Wisdom for All Ages from a Beloved Neighbor* (Viking, 1994), 115.

105 **The point is to name what's swirling:** You might wonder how this fits with research—like Ethan Kross's—that shows rehashing negative experiences can increase rumination and harm well-being. The key distinction is that rumination loops without resolution, while labeling a feeling—what psychologist Matthew Lieberman calls "affect labeling"—

helps the brain process it. Labeling activates the prefrontal cortex and calms the amygdala, making emotions more manageable. Rumination traps us in emotion; labeling gives us traction. Susan Nolen-Hoeksema, "The Role of Rumination in Depressive Disorders and Mixed Anxiety/ Depressive Symptoms," *Journal of Abnormal Psychology* 109, no. 3 (2): 504–11; Ethan Kross, Ozlem Ayduk, and Walter Mischel, "When Asking 'Why' Does Not Hurt: Distinguishing Rumination from Reflective Processing of Negative Emotions," *Psychological Science* 16, no. 9 (2005): 709–15; Jared B. Torre and Matthew D. Lieberman, "Putting Feelings into Words: Affect Labeling as Implicit Emotion Regulation," *Emotion Review* 10, no. 2 (March 2018): 116–24.

105 **UCLA psychologist Matthew Lieberman:** Katherina Kircanski, Matthew Lieberman, and Michelle G. Craske, "Feelings into Words: Contributions of Language to Exposure Therapy," *Psychological Science* 23, no. 10 (2012): 1086–91.

106 **Imaging studies have shown:** Matthew Lieberman, Naomi I. Eisenberger, Molly J. Crockett, Sabrina M. Tom, Jennifer H. Pfeifer, and Baldwin M. Way, "Putting Feelings into Words: Affect Labeling Disrupts Amygdala Activity in Response to Affective Stimuli," *Psychological Science* 18, no. 5 (2007): 421–28.

107 **He finally handed me:** The Wheel of Emotions was developed in the 1980s by psychologist Robert Plutchik. Since then, many different versions have been developed; this chapter includes one, which also draws on seminal work by psychologist James Russell. Robert Plutchik, "The Nature of Emotions: Human Emotions Have Deep Evolutionary Roots, a Fact That May Explain Their Complexity and Provide Tools for Clinical Practice," *American Scientist* 89, no. 4 (2001): 344–50; James A. Russell, "A Circumplex Model of Affect," *Journal of Personality and Social Psychology* 39, no. 6 (1980): 1161–78.

108 **Armed with a better understanding:** Nellie Duquette to Patricia and Bert Duquette, letter, June 22, 1943, North Bay, Ontario, Duquette Family Papers, private collection.

109 **As emotions scholar Amit:** Amit Goldenberg, conversation with author, 2025.

109 **Talk therapy has evolved:** Josef Breuer and Sigmund Freud, *Studies on Hysteria*, trans. James Strachey (Basic Books, 1957). Fun fact: Freud is

my academic great-great-grandfather—he's the great-grandfather of my PhD adviser, George (Freud) Loewenstein.

109 **If you've looked into it:** Aaron T. Beck, *Cognitive Therapy and the Emotional Disorders* (Penguin, 1976).

109 **It's been tested again:** Stefan G. Hofmann, Anu Asnaani, Imke J. J. Vonk, Alice T. Sawyer, and Angela Fang, "The Efficacy of Cognitive Behavioral Therapy: A Review of Meta-Analyses," *Cognitive Therapy and Research* 36, no. 5 (2012): 427–40; Stefan G. Hofmann and Jasper A. J. Smits, "Cognitive-Behavioral Therapy for Adult Anxiety Disorders: A Meta-Analysis of Randomized Placebo-Controlled Trials," *Journal of Clinical Psychiatry* 69, no. 4 (2008): 621–32.

109 **It's especially good:** Falk Leichsenring, Christiane Steinert, Sven Rabung, and John P. A. Ioannidis, "The Efficacy of Psychotherapies and Pharmacotherapies for Mental Disorders in Adults: An Umbrella Review and Meta-Analytic Evaluation of Recent Meta-Analyses," *World Psychiatry* 21, no. 1 (2022): 133–45; Pim Cuijpers, Hisashi Noma, Eirini Karyotaki, Christiaan H. Vinkers, Andrea Cipriani, and Toshi A. Furukawa, "A Network Meta-Analysis of the Effects of Psychotherapies, Pharmacotherapies and Their Combination in the Treatment of Adult Depression," *World Psychiatry* 19, no. 1 (February 2020): 92–107; Evan Mayo-Wilson, Sofia Dias, Ifigeneia Mavranezouli, Kayleigh Kew, David M. Clark, A. E. Ades, and Stephen Pilling, "Psychological and Pharmacological Interventions for Social Anxiety Disorder in Adults: A Systematic Review and Network Meta-Analysis," *The Lancet Psychiatry* 1, no. 5 (October 2014): 368–76; Lars-Göran Öst, Audun Havnen, Bjarne Hansen, and Gerd Kvale, "Cognitive Behavioral Treatments of Obsessive-Compulsive Disorder: A Systematic Review and Meta-Analysis of Studies Published 1993–2014," *Clinical Psychology Review* 40 (August 2015): 156–69.

109 **In fact, most evidence-based approaches:** Mary Lee Smith and Gene V. Glass, "Meta-Analysis of Psychotherapy Outcome Studies," *American Psychologist* 32, no. 9 (1977): 752–60; Bruce E. Wampold and Zac E. Imel, *The Great Psychotherapy Debate: The Evidence for What Makes Psychotherapy Work*, 2nd ed. (Routledge, 2015); Bruce E. Wampold, "How Important Are the Common Factors in Psychotherapy? An Update," *World Psychiatry* 14, no. 3 (2015): 270–77.

109 **This universality is known:** Lewis Carroll, *Alice's Adventures in Wonderland* (1865; reprint, Penguin Classics, 2003); Saul Rosenzweig, "Some Implicit Common Factors in Diverse Methods of Psychotherapy," *American Journal of Orthopsychiatry* 6, no. 3 (1936): 412–15.

110 **Just as the Dodo declared:** Carroll, *Alice's Adventures in Wonderland*, chapter 3.

110 **Friends and loved ones:** Chad E. Shenk and Alan E. Fruzzetti, "The Impact of Validating and Invalidating Responses on Emotional Reactivity," *Journal of Social and Clinical Psychology* 30, no. 2 (2011): 163–83; Marsha M. Linehan, "Validation and Psychotherapy," in *Empathy Reconsidered: New Directions in Psychotherapy*, ed. Arthur C. Bohart and Leslie S. Greenberg (American Psychological Association, 1997), 353–92.

110 **This process is called:** Frédéric Nils and Bernard Rimé, "Beyond the Myth of Venting: Social Sharing Modes Determine the Benefits of Emotional Disclosure," *European Journal of Social Psychology* 42, no. 6 (2012): 672–81; Stephen J. Lepore, Pablo Fernández-Berrocal, John Ragan, and Natalia Ramos, "It's Not That Bad: Social Challenges to Emotional Disclosure Enhance Adjustment to Stress," *Anxiety, Stress & Coping: An International Journal* 17, no. 4 (2004): 341–61.

110 **Here's some evidence of its benefit:** Razia S. Sahi, Elizabeth M. Gaines, Siyan G. Nussbaum, Daniel Lee, Matthew D. Lieberman, Naomi I. Eisenberger, and Jennifer A. Silvers, "You Changed My Mind: Immediate and Enduring Impacts of Social Emotion Regulation," *Emotion* 25, no. 2 (March 2025): 330–39.

111 **Speaking of which, there are different:** Razia S. Sahi, Zhouzhou He, Jennifer A. Silvers, and Naomi I. Eisenberger, "One Size Does Not Fit All: Decomposing the Implementation and Differential Benefits of Social Emotion Regulation Strategies," *Emotion* 23, no. 6 (September 2023): 1522–35.

111 **That's led some, myself included:** Michael V. Heinz et al., "Randomized Trial of a Generative AI Chatbot for Mental Health Treatment," *NEJM AI* 2, no. 4 (2025).

111 **It turns out that even generic:** Jamil Zaki and W. Craig Williams, "Interpersonal Emotion Regulation," *Emotion* 13, no. 5 (2013): 803–10; Razia S. Sahi, Emilia Ninova, and Jennifer A. Silvers, "With a Little Help from My Friends: Selective Social Potentiation of Emotion

Regulation," *Journal of Experimental Psychology: General* 150, no. 6 (2021): 1237–49.

111 **But it's not without risks:** Aishik Mandal, Tanmoy Chakraborty, and Iryna Gurevych, "Towards Privacy-Aware Mental Health AI Models: Advances, Challenges, and Opportunities," arXiv, February 1, 2025, doi.org/10.48550/arXiv.2502.00451; Kit Huckvale, John Torous, and Mark E. Larsen, "Assessment of the Data Sharing and Privacy Practices of Smartphone Apps for Depression and Smoking Cessation," *JAMA Network Open* 2, no. 4 (2019): e192542.

112 **showed participants three videos:** Olivier Luminet, Patrick Bouts, Frédérique Delie, Antony S. R. Manstead, and Bernard Rimé, "Social Sharing of Emotion Following Exposure to a Negatively Valenced Situation," *Cognition and Emotion* 14, no. 5 (2): 661–88.

112 **They concluded that the urge:** Bernard Rimé, "Emotion Elicits the Social Sharing of Emotion: Theory and Empirical Review," *Emotion Review* 1, no. 1 (2009), 60–85.

112 **In a study I conducted:** Mario L. Small, Kristina P. Brant, Ximena Garcia-Rada, and Leslie K. John, "People Avoid Their Support Network as Much as They Approach It—to Their Own Detriment" (working paper, Harvard Business School).

113 **His substance of choice? LSD:** Marty Jourard, *Sidney Jourard and the Transparent Self: The Humanistic Vision of a Pioneer Psychologist* (Book-Baby, 2021).

113 **Once, he used a projective test:** Julian B. Rotter and Janet E. Rafferty, *Manual for the Rotter Incomplete Sentences Blank* (The Psychological Corporation, 1950).

115 **That year, Anton Köllisch:** Roland W. Freudenmann, Florian Öxler, and Sabine Bernschneider-Reif, "The Origin of MDMA (Ecstasy) Revisited: The True Story Reconstructed from the Original Documents," *Addiction* 101, no. 9 (2006): 1241–45.

115 **Decades later, during the Cold War:** Terry Gross, interview with Stephen Kinzer, "The CIA's Secret Quest for Mind Control: Torture, LSD and a 'Poisoner in Chief,'" *Fresh Air*, NPR, September 9, 2019, audio, npr.org/2019/09/09/758989641/the-cias-secret-quest-for-mind-control-torture-lsd-and-a-poisoner-in-chief.

115 **So MDMA was shelved again:** James Hallifax, "A Brief History of MDMA: From the CIA to Raves to Psychedelic Therapy," *Psychedelic Spotlight*, May 16, 2022, psychedelicspotlight.com/history-of-mdma-cia-raves-psychedelic-therapy.

115 **It wasn't until the 1970s:** Drake Bennett, "Dr. Ecstasy," *New York Times*, January 30, 2005, nytimes.com/2005/01/30/magazine/dr-ecstasy.html; Alexander T. Shulgin and Ann Shulgin, *PiHKAL: A Chemical Love Story* (Transform Press, 1991).

115 **Rats given MDMA exhibit:** M. R. Thompson, Paul D. Callaghan, Glenn E. Hunt, Jennifer L. Cornish, and Iain S. McGregor, "A Role for Oxytocin and 5-HT1A Receptors in the Prosocial Effects of 3,4-Methylenedioxymethamphetamine ('Ecstasy')," *Neuroscience* 146, no. 2 (2007): 509–14.

115 **In primates, MDMA increases:** Sébastien Ballesta, Gilles Reymond, Matthieu Pozzobon, Jean-René Duhamel, "Effects of MDMA Injections on the Behavior of Socially-Housed Long-Tailed Macaques (*Macaca fascicularis*)," *PLOS ONE* 11, no. 2 (2016): e0147136.

116 **But maybe the most intriguing:** Eric Edsinger and Gül Dölen, "A Conserved Role for Serotonergic Neurotransmission in Mediating Social Behavior in Octopus," *Current Biology* 28, no. 19 (2018): 3136–42.e4.

116 **In 2021, a landmark double-blind:** Jennifer M. Mitchell, Michael Bogenschutz, Alia Lilienstein, et al., "MDMA-Assisted Therapy for Severe PTSD: A Randomized, Double-Blind, Placebo-Controlled Phase 3 Study," *Nature Medicine* 27 (2021): 1025–33.

116 **Though these results seem strong:** Suresh D. Muthukumaraswamy, Anna Forsyth, and Rachael L. Sumner, "The Challenges Ahead for Psychedelic 'Medicine,'" *Australian & New Zealand Journal of Psychiatry* 56, no. 11 (2022): 1378–83; Will Stone, "FDA Advisers Reject MDMA Therapy for PTSD, amid Concerns over Research," NPR, June 4, 2024, npr.org/2024/06/04/1252716275/fda-mdma-therapy-ptsd.

116 **That said, the promise:** Vicka Rael Corey, Vincent D. Pisano, and John H. Halpern, "Effects of 3,4-Methylenedioxymethamphetamine on Patient Utterances in a Psychotherapeutic Setting," *Journal of Nervous and Mental Disease* 204, no. 7 (July 2016): 519–23; Matthew J. Baggott, Matthew G. Kirkpatrick, Gillinder Bedi, and Harriet de Wit,

"Intimate Insight: MDMA Changes How People Talk about Significant Others," *Journal of Psychopharmacology* 29, no. 6 (June 2015): 669–77; Matthew J. Baggott et al., "Effects of 3,4-Methylenedioxymethamphetamine on Socioemotional Feelings, Authenticity, and Autobiographical Disclosure in Healthy Volunteers in a Controlled Setting," *Journal of Psychopharmacology* 30, no. 4 (April 2016): 378–87.

116 **Indeed, many MDMA trial participants:** Lester Grinspoon and James B. Bakalar, "Can Drugs Be Used to Enhance the Psychotherapeutic Process?," *American Journal of Psychotherapy* 40, no. 3 (1986): 393–404.

116 **It also floods the brain:** Danilo De Gregorio, Argel Aguilar-Valles, Katrin H. Preller, Boris Dov Heifets, Meghan Hibicke, Jennifer Mitchell, and Gabriella Gobbi, "Hallucinogens in Mental Health: Preclinical and Clinical Studies on LSD, Psilocybin, MDMA, and Ketamine," *Journal of Neuroscience* 41, no. 5 (2021): 891–900.

116 **Some researchers even speculate:** Margaret C. Wardle and Harriet de Wit, "MDMA Alters Emotional Processing and Facilitates Positive Social Interaction," *Psychopharmacology (Berlin)* 231, no. 21 (October 2014): 4219–29.

117 **In one study, Diana Tamir:** Diana I. Tamir and Jason P. Mitchell, "Disclosing Information about the Self Is Intrinsically Rewarding," *Proceedings of the National Academies of Sciences* 109, no. 21 (May 7, 2012): 8038–43.

117 **We like sharing so much:** Tamir and Mitchell, "Disclosing Information."

117 **Tamir and Mitchell found that people:** Tamir and Mitchell, "Disclosing Information."

118 **There were the flight attendants:** Ben Quinn, "Virgin Sacks 13 over Facebook 'Chav' Remarks, *Guardian*, October 31, 2008, theguardian.com/business/2008/nov/01/virgin-atlantic-facebook; Laura Bennett, "The First-Person Industrial Complex," *Slate*, September 14, 2015, slate.com/articles/life/technology/2015/09/the_first_person_indus trial_complex_how_the_harrowing_personal_essay_took.html.

118 **Offline, too, the memoir boom:** Ben Yagoda, *Memoir: A History* (Riverhead, 2010).

118 **When I was in graduate school:** Note that the admission rates reported for the version that was stripped down and relatively professional-

looking correspond to the "baseline" interface from the journal article: Leslie K. John, Alessandro Acquisti, and George Loewenstein, "Strangers on a Plane: Context-Dependent Willingness to Divulge Sensitive Information," *Journal of Consumer Research* 37, no. 5 (February 2011): 858–73.

120 **That's because whenever I presented:** Leslie John, "Ever Suffered from Selfie Regret? Why Some People Share When They Shouldn't," filmed 2016 at WIRED Conference, YouTube video, 13:48, posted by WIRED UK, November 2, 2016, youtube.com/watch?v=5VAAUerbgDw.

7: Building Friendship

123 **Talking about the last time:** Alison Wood Brooks, conversation with author, 2025.

124 **Mutual self-disclosure is:** Robert B. Hays, "A Longitudinal Study of Friendship Development," *Journal of Personality and Social Psychology* 48, no. 4 (1985): 909–24.

124 **First, revealing signals:** Harry T. Reis and Phillip Shaver, "Intimacy as an Interpersonal Process," in *Handbook of Personal Relationships: Theory, Research and Interventions*, ed. Steve Duck, Dale F. Hay, Stevan E. Hobfoll, William Ickes, and B. M. Montgomery (John Wiley & Sons, 1988), 367–89.

124 **The second reason that revelation:** Arthur Aron, Elaine N. Aron, Meg Tudor, and Greg Nelson, "Close Relationships as Including Other in the Self," *Journal of Personality and Social Psychology* 60, no. 2 (1991): 241–53; Nancy L. Collins and Lynn C. Miller, "Self-Disclosure and Liking: A Meta-Analytic Review," *Psychological Bulletin* 116, no. 3 (1994): 457–75.

126 **Anthropologists and evolutionary psychologists:** Michael Gurven, "To Give and to Give Not: The Behavioral Ecology of Human Food Transfers," *Behavioral and Brain Sciences* 27, no. 4 (2004): 543–83; K. Hawkes, F. O'Connell, and N. G. Blurton Jones, "Hunting and Nuclear Families: Some Lessons from the Hazda about Men's Work," *Current Anthropology* 42, no. 5 (2001): 681–709; Richard Wrangham, *Catching Fire: How Cooking Made Us Human* (Basic Books, 2009).

126 **In an influential 1960 article:** Alvin W. Gouldner, "The Norm of

Reciprocity: A Preliminary Statement," *American Sociological Review* 25 (1960): 161–78.

126 **Zoologist Matt Ridley suggests:** Matt Ridley, *The Origins of Virtue: Human Instincts and the Evolution of Cooperation* (Viking, 1997).

126 **Paleoanthropologist Richard Leakey:** Richard Leakey and Roger Lewin, *People of the Lake: Mankind and Its Beginnings* (Anchor Press/ Doubleday, 1978).

128 **The person who takes the first:** Sidney M. Jourard, "Self-Disclosure and Other-Cathexis," *Journal of Abnormal and Social Psychology* 59, no. 3 (1959): 428–31; Howard J. Ehrlich and David B. Graeven, "Reciprocal Self-Disclosure in a Dyad," *Journal of Experimental Social Psychology* 7, no. 4 (1971): 389–400; Morgan Worthy, Albert L. Gary, and Gay M. Kahn, "Self-Disclosure as an Exchange Process," *Journal of Personality and Social Psychology* 13, no. 1 (1969): 59–63.

128 **thirty-six increasingly personal questions:** Arthur Aron, Edward Melinat, Elaine N. Aron, Robert Darrin Vallone, and Renee J. Bator, "The Experimental Generation of Interpersonal Closeness: A Procedure and Some Preliminary Findings," *Personality and Social Psychology Bulletin* 23, no. 4 (1997): 363–77; Constantine Sedikides, W. Keith Campbell, Glenn D. Reader, and Andrew J. Elliot, "The Relationship Closeness Induction Task," *Representative Research in Social Psychology* 23 (1999): 1–4.

128 **any kind of close relationship:** Daniel Jones, "The 36 Questions That Lead to Love," *New York Times*, January 9, 2015, nytimes.com/2015/01 /09/style/no-37-big-wedding-or-small.html; Aron et al., "The Experimental Generation of Interpersonal Closeness."

129 **In a variation of the experiment:** Susan Sprecher, Stanislav Treger, Joshua D. Wondra, Nicole Hilaire, and Kevin Wallpe, "Taking Turns: Reciprocal Self-Disclosure Promotes Liking in Initial Interactions," *Journal of Experimental Social Psychology* 49, no. 5 (September 2013): 860–66; Susan Sprecher and Stanislav Treger, "The Benefits of Turn-Taking Reciprocal Self-Disclosure in Get-Acquainted Interactions," *Personal Relationships* 22, no. 3 (2015): 460–75.

130 **What matters in these moments:** Myong Jin Won-Doornink, "Self-Disclosure and Reciprocity in Conversation: A Cross-National Study," *Social Psychology Quarterly* 48, no. 2 (1985): 97–107; Jean-Philippe Lau-

renceau, Lisa Feldman Barrett, and Paula R. Pietromonaco, "Intimacy as an Interpersonal Process: The Importance of Self-Disclosure, Partner Disclosure, and Perceived Partner Responsiveness in Interpersonal Exchanges," *Journal of Personality and Social Psychology* 74, no. 5 (1998): 1238–51.

130 **my treasured colleague Youngme Moon:** Youngme Moon, "Intimate Exchanges: Using Computers to Elicit Self-Disclosure from Consumers," *Journal of Consumer Research* 26, no. 4 (2000): 323–39.

132 **The most jarring kind:** Laurenceau, Barrett, and Pietromonaco, "Intimacy as an Interpersonal Process"; Reis and Shaver, "Intimacy as an Interpersonal Process."

133 **As psychologist Lee Ross:** Lee Ross, "The Intuitive Psychologist and His Shortcomings: Distortions in the Attribution Process," *Advances in Experimental Social Psychology* 10 (1977): 173–220. Ross built on work by Edward E. Jones and Victor A. Harris, "The Attribution of Attitudes," *Journal of Experimental Social Psychology* 3, no. 1 (1967): 1–24. A related concept is the correspondence bias; Daniel T. Gilbert and Patrick S. Malone, "The Correspondence Bias," *Psychological Bulletin* 117, no. 1 (1995): 21–38.

133 **At the other extreme:** Irwin Altman and Dalmas Arnold Taylor, *Social Penetration: The Development of Interpersonal Relationships* (Holt, Rinehart & Winston, 1973); Alan L. Chaikin and Valerian J. Derlega, "Liking for the Norm-Breaker in Self-Disclosure," *Journal of Personality* 42, no. 1 (1974): 117–29; Paul C. Cozby, "Self-Disclosure, Reciprocity, and Liking," *Sociometry* 35, no. 1 (1972): 151–60; Richard L. Archer and John H. Berg, "Disclosure Reciprocity and Its Limits: A Reactance Analysis," *Journal of Experimental Social Psychology* 14, no. 5 (1978): 527–40.

136 **In research led by Mike Yeomans:** Karen Huang, Michael Yeomans, Alison Wood Brooks, Julia Minson, and Francesca Gino, "It Doesn't Hurt to Ask: Question-Asking Increases Liking," *Journal of Personality and Social Psychology* 113, no. 3 (2017): 430–52.

136 **According to Einav Hart:** Einav Hart, Eric M. VanEpps, and Maurice E. Schweitzer, "The (Better Than Expected) Consequences of Asking Sensitive Questions," *Organizational Behavior and Human Decision Processes* 162 (2021): 136–54.

136 **It happens when we mistake:** Anuj K. Shah and Michael LaForest, "Knowledge about Others Reduces One's Own Sense of Anonymity," *Nature* 603, no. 7900 (2022): 297–301.

138 **This illusion thrives in *parasocial relationships*:** Donald Horton and R. Richard Wohl, "Mass Communication and Para-Social Interaction: Observations on Intimacy at a Distance," *Psychiatry* 19, no. 3 (1956): 215–29.

138 **A classic example:** "'Lady Trapped in Television Set' Has Millions of Fans—All Under 6," *Sunday Journal and Star*, July 29, 1956, 40, newspapers.com/article/sunday-journal-and-star-frances-horwich/15613647.

138 **We asked participants to list:** Jaewon Yoon and Leslie John, "The Illusion of Reciprocity," unpublished data, ca. 2017.

140 **Anthropologist Robin Dunbar:** Maria Konnikova, "The Limits of Friendship," *New Yorker*, October 7, 2014, newyorker.com/science/maria-konnikova/social-media-affect-math-dunbar-number-friendships.

140 **a few ride-or-die:** Sheon Han, "You Can Only Maintain So Many Close Friendships," *Atlantic*, May 20, 2021, theatlantic.com/family/archive/2021/05/robin-dunbar-explains-circles-friendship-dunbars-number/618931.

140 **In the closest relationships:** Emma M. Templeton, Luke J. Chang, Elizabeth A. Reynolds, Marie D. Cone LeBeaumont, and Thalia Wheatley, "Long Gaps between Turns Are Awkward for Strangers but Not for Friends," *Philosophical Transactions of the Royal Society B: Biological Sciences* 378, no. 1875 (March 6, 2023): 20210471; Netta Weinstein, Thuy-vy Nguyen, Mark Adams, and C. Raymond Knee, "Intimate Sounds of Silence: Its Motives and Consequences in Romantic Relationships," *Motivation and Emotion* 48 (2024): 295–320.

144 **In one study, a full 84 percent:** Annabelle R. Roberts, Emma E. Levine, and Ovul Sezer, "Hiding Success," *Journal of Personality and Social Psychology* 120, no. 5 (2021): 1261–86.

144 **Envy is what psychologists call:** Niels Van de Ven, Marcel Zeelenberg, and Rik Pieters, "Leveling Up and Down: The Experiences of Benign and Malicious Envy," *Emotion* 9, no. 3 (2009): 419–29.

144 **Because if and when your friend:** Roberts, Levine, and Sezer, "Hiding Success."

144 **But that tends to backfire:** Ovul Sezer, Francesca Gino, and Michael I. Norton, "Humblebragging: A Distinct—and Ineffective—Self-Presentation Strategy," *Journal of Personality and Social Psychology* 114, no. 1 (January 2018): 52–74.

145 **In Alison Wood Brooks's:** Alison Wood Brooks, Karen Huang, Nicole Abi-Esber, Ryan W. Buell, Laura Huang, and Brian Hall, "Mitigating Malicious Envy: Why Successful Individuals Should Reveal Their Failures," *Journal of Experimental Psychology: General* 148, no. 4 (2019): 667–87.

146 **A visiting scholar:** Gavin J. Kilduff, Hillary Anger Elfenbein, and Barry M. Staw, "The Psychology of Rivalry: A Relationally Dependent Analysis of Competition," *Academy of Management Journal* 53, no. 5 (2010): 943–69.

148 **Emma Templeton and colleagues:** Templeton et al., "Long Gaps between Turns."

148 **In his memoir *We Should Not*:** Will Schwalbe, *We Should Not Be Friends: The Story of a Friendship* (Knopf, 2023): 42–43.

148 **"After almost four decades":** Schwalbe, *We Should Not Be Friends*, 291.

8: Finding Love

150 **"You have the right to remain silent":** *Miranda v. Arizona*, 384 U.S. 436 (1966).

150 **One woman who uses a wheelchair:** Kaitlin Menza, "3 Rings, 2 College Students and 1 Big Risk," *New York Times*, August 13, 2024, nytimes.com/2024/08/09/style/arya-singh-logan-mundy-wedding.html.

152 **My colleague (and bestie!):** Michael I. Norton, Jeana H. Frost, and Dan Ariely, "Less Is More: The Lure of Ambiguity, or Why Familiarity Breeds Contempt," *Journal of Personality and Social Psychology* 92, no. 1 (January 2007): 97–105.

152 **Eli Finkel and colleagues note:** Natasha D. Tidwell, Paul W. Eastwick, and Eli J. Finkel, "Perceived, Not Actual, Similarity Predicts Initial Attraction in a Live Romantic Context: Evidence from the Speed-Dating Paradigm," *Personal Relationships* 20, no. 2 (2013): 199–215; Shanhong Luo and Eva C. Klohnen, "Assortative Mating and

Marital Quality in Newlyweds: A Couple-Centered Approach," *Journal of Personality and Social Psychology* 88, no. 2 (2005): 304–26.

152 **not just for personality traits:** R. Matthew Montoya, Robert S. Horton, and Jeffrey Kirchner, "Is Actual Similarity Necessary for Attraction? A Meta-Analysis of Actual and Perceived Similarity," *Journal of Social and Personal Relationships* 25, no. 6 (2008): 889–922.

152 **for physical attractiveness:** Alan Feingold, "Matching for Attractiveness in Romantic Partners and Same-Sex Friends: A Meta-Analysis and Theoretical Critique," *Psychological Bulletin* 104, no. 2 (1988): 226–35.

153 **Studies have shown a curvilinear:** Paul W. Eastwick, Eli J. Finkel, Dan Mochon, and Dan Ariely, "Selective versus Unselective Romantic Desire: Not All Reciprocity Is Created Equal," *Psychological Science* 18, no. 4 (2007): 317–19; Peter K. Jonason and Norman P. Li, "Playing Hard-to-Get: Manipulating One's Perceived Availability as a Mate," *European Journal of Personality* 27, no. 5 (2013): 458–69; Elaine Walster, G. William Walster, Jane Piliavin, and Lynn Schmidt, "'Playing Hard to Get': Understanding an Elusive Phenomenon," *Journal of Personality and Social Psychology* 26, no. 1 (1973): 113–21.

154 **In a 1971 study that would:** Elaine Walster, G. William Walster, and P. Lambert, "Playing Hard-to-Get: A Field Study," University of Wisconsin, Madison, unpublished manuscript, 1971. Described in Walster, Walster, Piliavin, and Schmidt, "'Playing Hard to Get.'"

154 **Yet this balance:** Juliana Schroeder and Ayelet Fishbach, "Feeling Known Reflects Relationship Satisfaction," *Journal of Experimental Social Psychology* 111 (March 2024): 1–15.

155 **When we're highly motivated:** Sandra L. Murray, John G. Holmes, and Dale W. Griffin, "The Benefits of Positive Illusions: Idealization and the Construction of Satisfaction in Close Relationships," *Journal of Personality and Social Psychology* 70, no. 1 (1996): 79–98; Garth J. O. Fletcher, Jeffry A. Simpson, Geoff Thomas, and Louise Giles, "Ideals in Intimate Relationships," *Journal of Personality and Social Psychology* 76, no. 1 (1999): 72–89; Eastwick et al., "Selective versus Unselective Romantic Desire"; Oscar Ybarra and Walter G. Stephan, "Misanthropic Person Memory," *Journal of Personality and Social Psychology* 70, no. 4 (1996): 691–700.

155 **People are really good at curating:** Eli J. Finkel, Paul W. Eastwick,

Benjamin R. Karney, Harry T. Reis, and Susan Sprecher, "Online Dating: A Critical Analysis from the Perspective of Psychological Science," *Psychological Science in the Public Interest* 13, no. 1 (2012): 3–66.

156 **"the lone payoff":** Norton, Frost, and Ariely, "Less Is More," 103.

158 **Research led by Coral Zheng:** Coral Zheng and David Stillwell, "Ghosted? The Impacts of Ghosting in Interpersonal Communication," *Academy of Management Proceedings* 2023, no. 1 (August 2023).

162 **In fact, we tend to:** Michael Kardas, Amit Kumar, and Nicholas Epley, "Overly Shallow? Miscalibrated Expectations Create a Barrier to Deeper Conversation," *Journal of Personality and Social Psychology* 122, no. 3 (March 2022): 367–98; Michael Kardas, Amit Kumar, and Nicholas Epley, "Let It Go: How Exaggerating the Reputational Costs of Revealing Negative Information Encourages Secrecy in Relationships," *Journal of Personality and Social Psychology* 126, no. 6 (2024): 1052–83.

162 **In one study, I asked:** Leslie John, unpublished data, ca. 2021.

164 **Indeed, sometimes the hardest part:** Douglas Stone, Bruce Patton, and Sheila Heen, *Difficult Conversations: How to Discuss What Matters Most* (Viking, 1999).

165 **In a *Seinfeld* episode:** *Seinfeld*, season 6, episode 23, "The Face Painter," Andy Ackerman, dir., aired on NBC May 11, 1995, seinfeldism.com /s06e23-the-face-painter.php?fbclid=IwY2xjawJFW4dleHRuA2FlbQ IxMAABHS6xppczbymJLiSPZo5QbIScKL1Vj4b3DBOweR5lc6Kc tr2QC2PGm—4AQ_aem_4VYx1GnuZiUs7ME0lQODcQ.

165 **After all, for our early ancestors:** Geoff MacDonald and Mark R. Leary, "Why Does Social Exclusion Hurt? The Relationship between Social and Physical Pain," *Psychological Bulletin* 131, no. 2 (2005): 202.

165 **Roxy Music was right:** Roxy Music, "Love Is the Drug," track 1 on *Siren*, Island Records, 1975, Spotify.

166 **Studies have shown that daters:** Arthur Aron, Donald G. Dutton, Elaine N. Aron, and Adrienne Iverson, "Experiences of Falling in Love," *Journal of Social and Personal Relationships* 6, no. 3 (1989): 243–57; Susan Sprecher, "Insiders' Perspectives on Reasons for Attraction to a Close Other," *Social Psychology Quarterly* 61, no. 4 (1998): 287–300; Eastwick et al., "Selective versus Unselective Romantic Desire."

167 **When he finally gets up:** *Seinfeld*, season 6, episode 23.

168 **But as any *Seinfeld* fan:** Fun fact: In a deleted scene, Siena tells George she loves him and they get engaged.

168 **Social psychologist Joshua Ackerman:** Joshua M. Ackerman, Vladas Griskevicius, and Norman P. Li, "Let's Get Serious: Communicating Commitment in Romantic Relationships," *Journal of Personality and Social Psychology* 100, no. 6 (2011): 1079–94.

168 **This jibes with the widespread:** Alan Feingold, "Gender Differences in Personality: A Meta-Analysis," *Psychological Bulletin* 116, no. 3 (1994): 429–56.

169 **From an evolutionary perspective:** David M. Buss and David P. Schmitt, "Sexual Strategies Theory: An Evolutionary Perspective on Human Mating," *Psychological Review* 100, no. 2 (1993): 204–32.

170 **It's worth noting that different:** Richard Wilkins and Elisabeth Gareis, "Emotion Expression and the Locution 'I Love You': A Cross-Cultural Study," *International Journal of Intercultural Relations* 30, no. 1 (2006): 51–75.

170 **In a classic *60 Minutes*:** "Tango Finlandia," J. Tiffin, producer, CBS, *60 Minutes*, aired February 7, 1993, youtube.com/watch?v=kCKwe_Dp9Eg; Wilkins and Gareis, "Emotion Expression."

174 **Relationships often falter:** Caryl E. Rusbult and Paul A. M. Van Lange, "Interdependence, Interaction, and Relationships," *Annual Review of Psychology* 54, no. 1 (2003): 351–75; John M. Gottman and Robert W. Levenson, "Marital Processes Predictive of Later Dissolution: Behavior, Physiology, and Health," *Journal of Personality and Social Psychology* 63, no. 2 (1992): 221–33.

174 **Writer and podcaster:** Cat Sims, "How My Husband and I Fell Back in Love," *Not So Smug Now*, June 12, 2019, notsosmugnow.com/blog/2019 /6/11/how-my-husband-and-i-fell-back-in-love.

174 **A 2018 survey:** G. Oscar Anderson and Colette E. Thayer, "Loneliness and Social Connections: A National Survey of Adults 45 and Older," AARP Research (2018), aarp.org/content/dam/aarp/research/surveys _statistics/life-leisure/2018/loneliness-social-connections-2018.doi.10 .26419-2Fres.00246.001.pdf.

175 **In one study, researchers videotaped:** Celine Hinnekens, Alan Sillars, Lesley L. Verhofstadt, and William Ickes, "Empathic Accuracy and

Cognitions during Conflict: An In-Depth Analysis of Understanding Scores," *Personal Relationships* 27, no. 1 (2020): 102–31.

175 **Many studies have assessed *empathic accuracy*:** William Ickes, *Empathic Accuracy* (Guilford Press, 1997), 2.

175 **For neutral, everyday conversations:** William Ickes, "Everyday Mind Reading Is Driven by Motives and Goals," *Psychological Inquiry* 22, no. 3 (2011): 200–206; Celine Hinnekens, Gaëlle Vanhee, Maarten De Schryver, William Ickes, and Lesley L. Verhofstadt, "Empathic Accuracy and Observed Demand Behavior in Couples," *Frontiers in Psychology* 7 (2016): article 1370; Geoff Thomas, Garth J. O. Fletcher, and Craig Lange, "On-Line Empathic Accuracy in Marital Interaction," *Journal of Personality and Social Psychology* 72, no. 4 (1997): 839–50; Lesley L. Verhofstadt, Ann Buysse, William Ickes, Mark Davis, and Inge Devoldre, "Support Provision in Marriage: The Role of Emotional Similarity and Empathic Accuracy," *Emotion* 8, no. 6 (2008): 792–802.

175 **Worse, most people:** Carol Marangoni, Stephen Garcia, William Ickes, and Gary Teng, "Empathic Accuracy in a Clinically Relevant Setting," *Journal of Personality and Social Psychology* 68, no. 5 (1995): 854–69; Kenneth Savitsky, Boaz Keysar, Nicholas Epley, Travis Carter, and Ashley Swanson, "The Closeness-Communication Bias: Increased Egocentrism among Friends versus Strangers," *Journal of Experimental Social Psychology* 47, no. 1 (2011): 269–73; Anu Realo, Jüri Allik, Aire Nõlvak, Raivo Valk, Tuuli Ruus, Monika Schmidt, and Tiina Eilola, "Mind-Reading Ability: Beliefs and Performance," *Journal of Research in Personality* 37, no. 5 (2003): 429–45.

176 **One study found that people:** Courtney N. Wright and Michael E. Roloff, "You Should *Just Know* Why I'm Upset: Expectancy Violation Theory and the Influence of Mind Reading Expectations (MRE) on Responses to Relational Problems," *Communication Research Reports* 32, no. 1 (2015): 10–19; Roy J. Eidelson and Norman Epstein, "Cognition and Relationship Maladjustment: Development of a Measure of Dysfunctional Relationship Beliefs," *Journal of Consulting and Clinical Psychology* 50, no. 5 (October 1982): 715–20. And in the following paper, distraught couples in therapy were more likely to have high mind-reading expectations: N. Epstein and R. J. Eidelson, "Unrealistic Beliefs

of Clinical Couples: Their Relationship to Expectations, Goals and Satisfaction," *American Journal of Family Therapy* 9, no. 4 (1981): 13–22.

176 **And when researchers developed:** Eidelson and Epstein, "Cognition and Relationship Maladjustment."

176 **Believing your partner should:** Daniel H. Baucom and Norman Epstein, *Cognitive-Behavioral Marital Therapy* (Brunner/Mazel, 1990); Daniel F. Barone, Philinda S. Hutchings, Heather J. Kimmel, Howard L. Traub, Joan T. Cooper, and Christine M. Marshall, "Increasing Empathic Accuracy through Practice and Feedback in a Clinical Interviewing Course," *Journal of Social and Clinical Psychology* 24, no. 2 (2005): 156–71.

176 **Researchers who have studied:** Jean-Philippe Laurenceau, Lisa Feldman Barrett, and Paula R. Pietromonaco, "Intimacy as an Interpersonal Process: The Importance of Self-Disclosure, Partner Disclosure, and Perceived Partner Responsiveness in Interpersonal Exchanges," *Journal of Personality and Social Psychology* 74, no. 5 (1998): 1238–51; Shelly L. Gable, Harry T. Reis, Emily A. Impett, and Evan R. Asher, "What Do You Do When Things Go Right? The Intrapersonal and Interpersonal Benefits of Sharing Positive Events," *Journal of Personality and Social Psychology* 87, no. 2 (2004): 228–45.

9: Workplace Revealing (and Concealing)

182 **There has been a lot of buzz:** Mike Robbins, *Bring Your Whole Self to Work: How Vulnerability Unlocks Creativity, Connection, and Performance* (Hay House, 2018).

183 **Our Canuck friend:** Erving Goffman, *The Presentation of Self in Everyday Life* (Anchor, 1959).

184 **NYU organizational scientist:** Julianna Pillemer, "Strategic Authenticity: Signaling Authenticity without Undermining Professional Image in Workplace Interactions," *Organization Science* 35, no. 5 (2024): 1641–59.

184 **In workplace contexts:** Pillemer calls this "strategic authenticity," emphasizing a deliberate balance between openness and professionalism. I prefer "discerning authenticity" to highlight thoughtful selectivity without the potentially manipulative connotation.

185 **Among the most-qualified applicants:** For weaker candidates, authenticity didn't help and sometimes hurt. The takeaway isn't "be inauthentic," but rather: Focus on roles where your strengths are a good match. Celia Moore, Sun Young Lee, Kawon Kim, and Daniel M. Cable, "The Advantage of Being Oneself: The Role of Applicant Self-Verification in Organizational Hiring Decisions," *Journal of Applied Psychology* 102, no. 11 (2017): 1493–1513.

186 **In one study, people who approached:** Daniel M. Cable and Virginia S. Kay, "Striving for Self-Verification during Organizational Entry," *Academy of Management Journal* 55, no. 2 (April 1, 2012): 360–80.

187 **Evolutionary biologists might:** Adrian Bangerter, Nicolas Roulin, and Cornelius J. König, "Personnel Selection as a Signaling Game," *Journal of Applied Psychology* 97, no. 4 (2012): 719–38.

187 **Up to 96 percent of candidates:** Julia Levashina and Michael A. Campion, "Measuring Faking in the Employment Interview: Development and Validation of an Interview Faking Behavior Scale," *Journal of Applied Psychology* 92, no. 6 (2007): 1638–56.

188 **"When organizations tell":** Alicia Menendez, "Women Are Often Told to Just Be Themselves. Here's Why That Advice Can Be Dangerous," *Time*, October 29, 2019, time.com/5704997/alicia-menendez-likeability-trap.

190 **As Kovacheff told me:** Chloe Kovacheff, interview with author, 2025.

190 **In general, research shows:** Stephenie R. Chaudoir and Jeffrey D. Fisher, "The Disclosure Processes Model: Understanding Disclosure Decision Making and Postdisclosure Outcomes among People Living with a Concealable Stigmatized Identity," *Psychological Bulletin* 136, no. 2 (2010): 236–56; John W. Lynch and Jessica Rodell, "Blend In or Stand Out? Interpersonal Outcomes of Managing Concealable Stigmas at Work," *Journal of Applied Psychology* 103, no. 12 (2018): 1307–23.

192 **Christine doesn't remember seeing the stars:** The information about Christine is based on my conversations with her in 2024, as well as: Christine Exley, "Share Your Vision: Christine Exley," Foundation Fighting Blindness, 2025, fightingblindness.org/shareyourvision-gallery-christine-exley.

193 **Christine's approach jibes:** Chloe Kovacheff, "Disconnected Disclosures: Employee-Manager Asymmetries in Navigating Invisible Disabilities" (PhD diss., University of Toronto, 2024); Chloe Kovacheff, interview with author.

193 **"If you're risk averse":** Chloe Kovacheff, interview with author.

195 **That stereotype can:** Laura M. Little, Virginia Smith Major, Amanda S. Hinojosa, and Debra L. Nelson, "Professional Image Maintenance: How Women Navigate Pregnancy in the Workplace," *Academy of Management Journal* 58, no. 1 (2015): 8–37.

195 **One qualitative study:** Laura Little, Amanda Hinojosa, and John Lynch, "Make Them Feel: How the Disclosure of Pregnancy to a Supervisor Leads to Changes in Perceived Supervisor Support," *Organization Science* 28, no. 4 (2017): 618–35.

196 **"You can't bring the baby":** Little et al., "Make Them Feel," 632.

196 **Craig was a leader:** The information about Craig is based on my interview with him on March 19, 2025, as well as: Craig Kramer, LinkedIn News, 2024, linkedin.com/feed/update/urn:li:activity:7089968663725764608.

200 **One survey of 3,078:** James Davis, "Have You Cried at Work?," *HR Daily Advisor*, October 21, 2019, hrdailyadvisor.blr.com/2019/10/21/have-you-cried-at-work.

201 **I didn't know it:** Elizabeth Baily Wolf, Jooa Julia Lee, Sunita Sah, and Alison Wood Brooks, "Managing Perceptions of Distress at Work: Reframing Emotion as Passion," *Organizational Behavior and Human Decision Processes* 137 (2016): 1–12.

202 **But the finding may:** Kimberly D. Elsbach and Beth A. Bechky, "How Observers Assess Women Who Cry in Professional Work Contexts," *Academy of Management Discoveries* 4, no. 2 (2018): 125–41; Victoria L. Brescoll, "Leading with Their Hearts? How Gender Stereotypes of Emotion Lead to Biased Evaluations of Female Leaders," *Leadership Quarterly* 27, no. 3 (2016): 415–28.

202 **One study of 1,500:** Melissa Morse, "Something to Cry About—25% of Employees Have Left a Performance Review Crying," *HR Daily Advisor*, January 30, 2017, hrdailyadvisor.blr.com/2017/01/30/something-cry-25-employees-left-performance-review-crying.

202 **As Olga Khazan concluded:** Olga Khazan, "Is It Okay to Cry at Work?," *Atlantic*, March 17, 2016, theatlantic.com/video/index/474195/is-it-okay-to-cry-at-work.

202 **According to writer Sandra Newman:** Sandra Newman, "Whatever Happened to the Noble Art of the Manly Weep?," *Aeon*, July 9, 2018, aeon.co/essays/whatever-happened-to-the-noble-art-of-the-manly-weep.

202 **Accounts of men weeping:** Tom Lutz, *Crying: A Natural and Cultural History of Tears* (W. W. Norton, 1999).

203 **How did we get away:** Newman, "Whatever Happened."

203 **They were trained to suppress:** Lutz, *Crying.*

203 **I hope that we:** Taylor Leamey, "The Benefits of Crying and Why It's Good for Your Health," CNET, June 5, 2022, cnet.com/health/medical/the-benefits-of-crying-and-why-its-good-for-your-health.

203 **oxytocin and endorphins:** Leo Newhouse, "Is Crying Good for You?," *Harvard Health Blog*, March 1, 2021, health.harvard.edu/blog/is-crying-good-for-you-2021030122020.

203 **parasympathetic nervous system:** MBT Desk, "The Importance of Crying: Physiology, Emotional Response, and Ocular Protection," *MedBound Times*, June 30, 2024, medboundtimes.com/medicine/importance-of-crying.

10: Why Great Leaders Share More

208 **"The last time I cried":** Diane Herbst, "Despite Praising Kavanaugh, Trump Doesn't Like When Men Cry: 'Not His Idea of Masculinity,'" *People*, September 28, 2018, people.com/politics/kavanaugh-tears-crying-trump-reaction.

208 **Herminia Ibarra, a professor:** Herminia Ibarra, "The Authenticity Paradox," *Harvard Business Review*, January 2015, hbr.org/2015/01/the-authenticity-paradox.

209 **What would Erving Goffman:** Erving Goffman, *The Presentation of Self in Everyday Life* (Anchor, 1959).

209 **Julianna Pillemer, whose work:** Julianna Pillemer, "Strategic Authenticity: Signaling Authenticity without Undermining Professional Image in Workplace Interactions," *Organization Science* 35, no. 5 (2024): 1647.

210 **Research by Susan Fiske:** Susan T. Fiske, Amy J. C. Cuddy, Peter Glick, and Jun Xu, "A Model of (Often Mixed) Stereotype Content: Competence and Warmth Respectively Follow from Perceived Status and Competition," *Journal of Personality and Social Psychology* 82, no. 6 (2002): 878–902.

211 **Women are often expected:** Katherine W. Phillips, Nancy P. Rothbard, and Tracy L. Dumas, "To Disclose or Not to Disclose? Status Distance and Self-Disclosure in Diverse Environments," *Academy of Management Review* 34, no. 4 (2009): 710–32.

212 **In one, we analyzed:** Leslie K. John, Martha Jeong, Francesca Gino, and Laura Huang, "The Self-Presentational Consequences of Upholding One's Stance in Spite of the Evidence," *Organizational Behavior and Human Decision Processes* 154 (2019): 1–14.

213 **In 2010, a newly minted:** Melanie Stefan, "A CV of Failures," *Nature* 468, no. 7322 (2010): 467.

213 **In a series of studies:** Li Jiang, Leslie K. John, Reihane Boghrati, and Maryam Kouchaki, "Fostering Perceptions of Authenticity via Sensitive Self-Disclosure," *Journal of Experimental Psychology: Applied* 28, no. 4 (2022): 898–915.

214 **After Zoom was rocked:** "Zoom CEO Addresses 'Zoombombing': We Had Some Missteps," CNN, April 5, 2020, youtube.com/watch?v=xk992LJ4N9M.

214 **His response was well received:** Casey Newton, "Zoom Faces a Privacy and Security Backlash as It Surges in Popularity," *The Verge*, April 1, 2020.

214 **As tech CEO:** Jim Whitehurst, "How to Earn Respect as a Leader," *Harvard Business Review*, May 20, 2015, hbr.org/2015/05/how-to-earn -respect-as-a-leader.

214 **A leader confident enough:** Amy C. Edmondson, *The Fearless Organization: Creating Psychological Safety in the Workplace for Learning, Innovation, and Growth* (Wiley, 2019).

215 **Well, in one study:** Constantinos Coutifaris and Adam M. Grant, "Taking Your Team behind the Curtain: The Effects of Leader Feedback-Sharing and Feedback-Seeking on Team Psychological Safety," *Organization Science, INFORMS* 33, no. 4 (July 2022), 1574–98.

215 **Disclosing unethical or immoral:** Annabelle Roberts, Emma E. Levine, and Justin F. Landy, "Disclosing Shortcomings in Morality, Sociability, and Competence: Differing Effects on Trust" (working paper).

215 **We looked again at leaders:** Jiang, John, Boghrati, and Kouchaki, supplemental study in "Fostering Perceptions of Authenticity."

215 **Even Cynthia, the healthcare executive:** Ibarra, "The Authenticity Paradox."

216 **Consider the now-infamous case:** "HAYWARD—LIFE BACK," CNN, May 30, 2010, www.youtube.com/watch?v=EZraCNZZ7U8.

216 **Under mounting pressure:** Clifford Krauss, "BP Plans to Replace Tony Hayward as Chief," *New York Times*, July 25, 2010, nytimes.com/2010 /07/26/business/global/26bp.html.

216 **Leadership expert Brené Brown:** Brené Brown, *Daring Greatly: How the Courage to Be Vulnerable Transforms the Way We Live, Love, Parent, and Lead* (Gotham Books, 2012).

217 **Consider, for example, how Queen Elizabeth:** Warren Hoge, "Responding to Britain's Sorrow, Queen Will Address the Nation," *New York Times*, September 5, 1997, nytimes.com/1997/09/05/world/re sponding-to-britain-s-sorrow-queen-will-address-the-nation.html.

217 **She addressed the nation:** Queen Elizabeth, "The Queen's Message Following the Death of Diana, Princess of Wales," September 5, 1997, royal.uk/queens-message-following-death-diana-princess-wales.

218 **As one onlooker:** "The Queen Did Something EXTRAORDINARY When Diana Died," The Royal Family Channel, YouTube video, 3:41, September 5, 2022, youtube.com/watch?v=7taE7jRhWR0.

218 **"She was being a proper granny":** Barry Neild and Nick Glass, "How Diana's Death Turned Queen into 'Proper Granny,'" CNN.com, June 5, 2012, edition.cnn.com/2012/05/30/world/europe/margaret-rhodes-queen /index.html.

218 **"It was not a quick bow":** Gillian Brockell, "The Moment the Queen Gave Princess Diana Her Due 25 Years Ago," *Washington Post*, September 10, 2022, washingtonpost.com/history/2022/09/10/queen-elizabeth -princess-diana-funeral.

218 **Years later, in 2005:** Carolyn Durand, "Letter from Queen Elizabeth

about Princess Diana's Death Comes to Light," ABC News, August 13, 2017, abcnews.go.com/International/letter-queen-elizabeth-princess-di ana-death-light/story?id=49188088.

219 **As Daniel Goleman, who popularized:** Daniel Goleman, *Emotional Intelligence: Why It Can Matter More Than IQ* (Bantam Books, 1995), 43.

219 **Instead, the better path:** Arlie Russell Hochschild, *The Managed Heart: Commercialization of Human Feeling* (University of California Press, 1983).

220 **As he spoke about:** Josh Lederman, "'It Gets Me Mad'—Obama Acts Alone on Gun Control," Associated Press, January 7, 2016, postinde pendent.com/news/local/it-gets-me-mad-obama-acts-alone-on-gun -control.

220 **"This is the most emotion":** John Blake, "Why Obama's Tears Are So Revolutionary," CNN.com, January 8, 2016, cnn.com/2016/01/08/pol itics/obama-gun-control-tears/index.html.

221 **Marriott International CEO Arne Sorenson:** Jason Aten, "Marriott's CEO Shared a Video with His Team and It's a Powerful Lesson in Leading during a Crisis," *Inc.*, March 20, 2020, inc.com/jason-aten /marriotts-ceo-shared-a-video-with-his-team-its-a-powerful-les son-in-leading-during-a-crisis.html.

221 **In a survey of two thousand:** L'Oreal Thompson Payton, "Crying at Work Is More than Okay, It's an Essential Part of Mental Health," Fortune Well, August 25, 2022, fortune.com/well/2022/08/25/crying-at -work-mental-health.

221 **Braden Wallake, CEO:** Braden Wallake, LinkedIn post, August 9, 2022, linkedin.com/feed/update/urn:li:activity:6962886723617910784 /?utm_source=vice&utm_medium=iframely.

222 **When *Huffington Post*:** Catherine Pearson, "What 15 Female Leaders Really Think about Crying at Work," *Huffington Post*, May 28, 2014, huffpost.com/entry/crying-at-work-women_n_5365872.

222 **These anecdotes aren't:** Elizabeth Wolf, Jooa Julia Lee, Sunita Sah, and Alison Wood Brooks, "Managing Perceptions of Distress at Work: Reframing Emotion as Passion," *Organizational Behavior and Human Decision Processes* 137, no. 1 (2016): 1–12.

223 **On November 7, 1991:** "The Announcement: Magic Johnson," National Basketball Association, youtube.com/watch?v=xMMWLS8D4OU.

224 **One study estimated:** Alexander Cardazzi, Joshua C. Martin, and Zachary Rodriguez, *Information Avoidance and Celebrity Exposure: The Effect of "Magic" Johnson on AIDS Diagnoses and Mortality in the U.S.*, Economics Faculty Working Papers Series, no. 57 (2021), research repository.wvu.edu/econ_working-papers/57.

225 **I recently caught up with Brook:** Brook Mahealani Lee, interview with author, March 26, 2025.

Epilogue

228 **Both speak to a fundamental:** Roy F. Baumeister and Mark R. Leary, "The Need to Belong: Desire for Interpersonal Attachments as a Fundamental Human Motivation," *Psychological Bulletin* 117, no. 3 (1995): 497–529; Martin Buber, *I and Thou*, trans. Ronald Gregor Smith (Charles Scribner's Sons, 1958); Dan P. McAdams, "The Psychology of Life Stories," *Review of General Psychology* 5, no. 2 (2001): 100–122.

228 **The craving to be authentically:** William B. Swann Jr., "Self-Verification: Bringing Social Reality into Harmony with the Self," in *Psychological Perspectives on the Self*, vol. 2, ed. Jerry Suls (Lawrence Erlbaum Associates, 1983), 33–66.

229 **For eight years, Australian palliative:** Bronnie Ware, *The Top Five Regrets of the Dying: A Life Transformed by the Dearly Departing* (Hay House, 2019).

229 **Flash back to chapter 3:** Thomas Gilovich and Victoria Husted Medvec, "The Temporal Pattern to the Experience of Regret," *Journal of Personality and Social Psychology* 67, no. 3 (1994): 357–65.

232 **For journalist Laura:** Laura Trujillo, *Stepping Back from the Ledge: A Daughter's Search for Truth and Renewal* (Random House, 2022).

232 **Crying, she told her children:** Trujillo, *Stepping Back*, 68.

232 **Handwritten on an index card:** Trujillo, *Stepping Back*, 70.

232 **"I carried it in my wallet":** Laura Trujillo, interview with author, November 1, 2024.

232 **She has also quite consciously:** Nicole Carroll, "You Can Talk about Suicidal Thoughts and Depression. USA TODAY Editor Shares Advice

after Her Mother's Death by Suicide," *USA TODAY*, April 22, 2022, fosters.com/story/opinion/2022/04/22/laura-trujillo-memoir-advice-depression-suicidal-thoughts/7394834001.

233 **Consider one of palliative care:** Ware, *The Top Five Regrets of the Dying*.

233 **Later, Jude's words stayed:** Ware, *The Top Five Regrets of the Dying*, 146.

Index

About the Author

Leslie John is a behavioural scientist and leading privacy expert at Harvard Business School whose research focuses on how we make decisions. She has been published in leading academic journals and has influenced scholars, companies, NGOs and governments around the world.

She's worked with organisations including Facebook, Goldman Sachs, J.P. Morgan and the Commonwealth Bank of Australia. Her research has also appeared in top media including the *New York Times*, the *Wall Street Journal*, *Financial Times*, and *Time Magazine*.

A Canadian-born internationally trained ballet dancer, she now calls Boston home.